essential science PHYSICS

P E Bloomfield
R Saul
S C Thompson

Oxford University Press

Oxford University Press, Walton Street, Oxford OX2 6DP

Oxford London Glasgow
New York Toronto Melbourne Auckland
Petaling Jaya Singapore Hong Kong Tokyo
Delhi Bombay Calcutta Madras Karachi
Nairobi Dar es Salaam Cape Town

and associated companies in
Beirut Berlin Ibadan Nicosia

OXFORD is a trade mark of Oxford University Press

© P. E. Bloomfield, R. Saul, S. C. Thompson 1984

First published 1984
Reprinted 1987

ISBN 0 19 914094 4

Photoset by Rowland Phototypesetting Ltd., Bury St Edmunds, Suffolk
Printed in Great Britain by Thomson Litho Ltd, East Kilbride, Scotland

Contents

1 Properties of matter
1.1 Units 1
1.2 Mass and weight 2
1.3 Density and volume 2
1.4 Pressure 4
1.5 Archimedes' principle 7
1.6 Extension of a spring 9

2 Mechanics
2.1 Moments 10
2.2 Centre of gravity 11
2.3 Vector and scalar addition 13
2.4 Speed, velocity and acceleration 15
2.5 Newton's Laws of Motion 18
2.6 Motion in a curved path 20
2.7 Force, work, energy and power 21
2.8 Machines 22

3 Wave motion and sound
3.1 Waves 24
3.2 Sound 27
3.3 Vibrating strings 29
3.4 The electromagnetic spectrum 30

4 Light
4.1 Light travels in straight lines 32
4.2 Reflection 34
4.3 Refraction 36
4.4 Lenses 39
4.5 The eye 41
4.6 Optical instruments 43
4.7 Colour 46

5 Heat
5.1 Molecules and the kinetic theory 48
5.2 Thermometry 50
5.3 Expansion of solids and liquids 52
5.4 The gas laws 55
5.5 Heat measurement 58
5.6 Change of state 60
5.7 Heat transmission 63

6 Magnetism and electricity
6.1 Magnetism 66
6.2 Electrostatics 69
6.3 Current and resistance 72
6.4 Electrical energy 76
6.5 Electromagnetism 79
6.6 Electromagnetic induction 82

7 Physics of the atom
7.1 The electron 86
7.2 The nucleus 89

Examination questions 93

Answers 113

Index 119

Acknowledgements 121

Preface

This book will satisfy the essential needs of students taking GCSE physics courses. It is designed not only for class work and homework use by students of various abilities but also for revision purposes.

The main aim of the book is to present a concise and straightforward account of the information with the minimum of general background. The seven chapters cover each of the major areas of the subject and are subdivided into short units for ease of reference. These contain the essential principles that have to be learned and details of the key experiments together with a number of worked examples to help in numerical practice. Each unit is followed by exercises of increasing difficulty for students to attempt immediately after studying the unit. In addition, questions from examination papers are included at the end of the book to give students an idea of the standards expected. This is followed by a section giving the numerical answers to all questions in the book.

Wherever appropriate, SI units (Le Système International d'Unités) and the recommendations of the Association for Science Education have been used although more common terms are also included in places.

We wish to thank the following examination boards for their permission to use questions from their papers: Associated Examining Board, Joint Matriculation Board, Oxford Delegacy of Local Examinations, Southern Regional Examination Board, University of London Entrance and Schools Examination Council, and Yorkshire and Humberside Regional Examinations Board.

Finally, we would like to thank all those concerned for their patient help and hard work on this book and for their support and encouragement.

P.E.B.
R.S.
S.C.T.

1 Properties of matter

1.1 Units

Units used in science are internationally agreed and are part of the Système Internationale d'Unités (the SI system).

Examples of SI units

physical quantity	name of SI unit	symbol for unit
length	metre	m
mass	kilogram	kg
time	second	s
temperature	kelvin	K
electric current	ampere	A
potential difference	volt	V
energy	joule	J
work	joule	J
power	watt	W
force	newton	N

Note: the kilogram is the SI unit of mass, and not the gram.

Writing the units

a) Write the words in full or use the agreed abbreviation, e.g. metres or m.
b) Never add an 's' to a symbol to make it plural, e.g. 5 kilograms = 5 kg.
c) A symbol is not given a capital letter unless it commemorates a person. When writing in full, do not use a capital letter, e.g. 5 newtons or 5 N, 10 watts or 10 W.
d) Do not put a full stop after a symbol except when it occurs at the end of a sentence.

Multiples of basic units

A prefix is used to denote the multiple of a unit, e.g. 1 kilometre = 1000 metres.
Here are some other important prefixes given for the metre as an example.

prefix	multiple	symbol	size compared to a metre
mega-	megametre	Mm	1 000 000 m
kilo-	kilometre	km	1000 m
	metre	m	1 m
milli-	millimetre	mm	0.001 m
micro-	micrometre	µm	0.000 001 m
nano-	nanometre	nm	0.000 000 001 m

Writing the numbers

a) When numbers are written in figures, no commas are used to separate the thousands. For numbers greater than 9999, the figures are grouped in three's, e.g.

    five and a half million   5 500 000
    five hundred thousand   500 000
    seven thousand two hundred and twelve   7212

Very small numbers are written out in a similar fashion, e.g. 0.0004, 0.000 019, 0.000 000 07.

b) Decimal points are placed on the line, e.g. 5.2, 0.5 and 72.34.
c) Standard index form is convenient for very large or very small numbers, e.g.
5 500 000 can be expressed as 5.5×10^6
0.000 000 012 becomes 1.2×10^{-8}

Exercises

1 Write the units of a) mass b) length c) time

2 Write these quantities in words: a) 6 km b) 7 s c) 2.5 kg d) 26 ms e) 7.5 mm f) 3 g g) 4 m h) 5.6 ks i) 68 mg

3 Write these quantities in numbers and symbols: a) seven millimetres b) twenty-five milligrams c) thirty-two kilometres d) three seconds e) twelve metres f) eighteen kilograms

4 Write the following in kilograms: a) 7000 g b) 3500 g c) 8250 g d) 9872 g e) 10001 g

5 Write the following in millimetres: a) 7 m b) 57.4 m c) 69 km

6 Write how many milliseconds there are in: a) 5 seconds b) 2 minutes c) 1 hour

7 If 1 km = 0.625 miles, how many kilometres are there in 30 miles? Express the 30 m.p.h. traffic speed limit in metres per second.

8 A ribbon 5.65 m long is divided equally between five pupils. How long is the piece that each receives?

9 Express the following numbers in index form: a) 3942 b) 27 000 000 000 c) 11 d) 0.001 e) 0.000 000 000 725

10 Express the following measurements in non-index form:
a) 2.6×10^6 m   b) 0.3×10^3 g   c) 0.1×10^{-1} s
d) 0.1×10^{-7} s   e) 26.5×10^{-11} N

1.2 Mass and weight

The *mass* of an object is the amount of matter that it contains.
The *weight* of an object is the force of gravity on it.

Gravity and weight All masses attract each other with the force of gravity. Masses m_A and m_B, separated by a distance r, experience a force of attraction F, so that

$$F = \frac{m_A \times m_B}{r^2} \cdot G$$

where G is a constant.
m_A and m_B are pulled by the same force.
All things on the Earth are attracted to the Earth by gravity. Part of the gravitational force is used to keep everything spinning through space with the Earth, and the remainder holds everything to the Earth. The force holding each object to the Earth is the object's weight.

Weight is a *force* and has all the properties of forces given on page 13.

Variation of mass and weight Under normal conditions, the mass of an object never changes. Its weight varies very slightly as it is moved over the surface of the Earth because the Earth is not a perfect sphere. An astronaut on the Moon has the same mass that he has on Earth. His weight is much less however because of the lower gravity experienced.

Units of mass and weight

Mass is measured in kilograms (kg)
Weight is measured in newtons (N)
A mass of 1 kg weighs approximately 10 N.

Measurement of mass and weight

Weight is measured with a spring balance.
Mass is measured using a beam balance and standard masses.

Exercises

1 How much do these masses weigh? a) 5 kg   b) 7.25 kg   c) 9.5 kg   d) 3.2 kg   e) 0.8 kg   f) 0.02 kg   g) 0.005 kg   h) 0.004 kg

2 Write down the mass of these weights. a) 10 N   b) 5 N   c) 20 N   d) 32 N   e) 0.5 N   f) 0.05 N   g) 0.58 N   h) 0.003 N

3 What instrument would you choose to measure a) mass and b) weight? Briefly explain how each instrument is used.

4 Why does the weight of an object vary over the Earth's surface? If the weight varies, why does the mass not vary as well?

1.3 Density and volume

The *volume* of an object is the space that it occupies. The usual units of volume are the cubic centimetre (cm³) and the cubic metre (m³).
1 m³ = 100 × 100 × 100 cm³ = 1 000 000 cm³
= 1 × 10⁶ cm³

Calculating volumes

The volume of a regularly shaped object can be found by calculation.

a) The cube
   volume = $a \times a \times a$
   = a^3

b) The rectangular block
   volume = $p \times q \times s$

c) The sphere
   volume = $\frac{4}{3}\pi r^3$

d) The cylinder
   volume = (area of cross section) × height
   = $\pi r^2 h$

Volume measurement

The volume of an irregularly shaped object is found from the displacement of liquid in a measuring cylinder. Any object totally immersed in a liquid will displace a volume of liquid equal to its own volume.

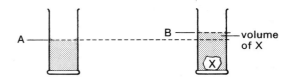

The level of liquid is read with the eye level with the bottom of the meniscus. The difference between the levels of liquid before and after immersion of the object gives the volume of the object. If the irregularly shaped object is too large to go into the measuring cylinder, the liquid may be displaced from a displacement (eureka) can into the cylinder as shown below.

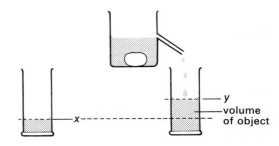

It is slightly more difficult to find the volume of an object that floats in the liquid. The object is made to sink by attaching it to a heavy mass. The volume of liquid thus displaced is measured in a cylinder. The volume of the heavy mass must be subtracted from this result to give the required volume of the lighter object.

Density

Equal volumes of different substances do not have the same mass.

The density of a substance is the mass, in kilograms, of a cubic metre of the substance.

The units of density are kg/m^3, although it is often expressed in g/cm^3. Notice that $1\ g/cm^3 = 1000\ kg/m^3$.

Measurement of density of a solid

It is not necessary to have a cubic metre of a substance in order to determine its density. Simply take any small amount of the solid and

a) find its mass with a beam balance
b) find its volume using a measuring cylinder and (if necessary) a displacement can.

Then, density $= \dfrac{\text{mass}}{\text{volume}}$

Measuring the density of a liquid

Use a density bottle. A density bottle contains a definite volume of liquid when full, e.g. $25\ cm^3$. The actual volume is usually engraved on the side of the bottle. A ground glass stopper with a fine hole through it fits into the neck. The bottle should be filled, the stopper put in place, and then the top wiped with filter paper. The density bottle then contains a definite volume of liquid.

Subtract the mass of the clean dry bottle from the mass of the bottle when filled with liquid. Then apply the density equation:

density $= \dfrac{\text{mass}}{\text{volume}}$

Relative density of a liquid

A density bottle is used to find the relative density of a liquid where

relative density of a liquid $= \dfrac{\text{mass of any volume of a liquid}}{\text{mass of equal volume of water}}$

Note: relative density has no units.

Exercises

1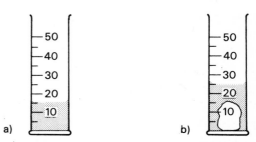

Fig a) shows water in a measuring cylinder. Fig b) shows the same measuring cylinder and water with a stone in it. What is the volume of the stone if the water is measured in cm^3?

2 Calculate the volumes of these objects.
 a) a container $7\ m \times 2\ m \times 4\ m$
 b) a cube of side $2\ m$
 c) a sphere of radius $6\ m$
 d) a cylinder $8\ m$ high and diameter $5\ m$

3 Write down these volumes in cm^3: a) $0.75\ m^3$
b) $0.085\ m^3$ c) $6.75\ m^3$

4 Write down these volumes in m^3: a) $7\,000\,000\ cm^3$
b) $650\,000\ cm^3$ c) $750\ cm^3$

5 The table shows some of the measurements of a set of rectangular blocks. Copy and complete the table.

length (m)	breadth (m)	height (m)	volume (m^3)
2	6	3	
	4	2	24
20		10	1000
9	3		5.4

6 Copy and complete this table.

mass (g)	volume (cm^3)	density (g/cm^3)
48	6	
	2	3
56		14

7 Use the table of densities below to answer these questions.
 a) What is the mass of $3\ cm^3$ of silver?
 b) What is the mass of a mixture of $1000\ cm^3$ of water and $200\ cm^3$ of lead shot?
 c) What volume does $10\ kg$ of mercury occupy?
 d) Find the volume of $75.25\ g$ of platinum.

substance	density (kg/m^3)	density (g/cm^3)
silver	10 500	10.5
lead	11 300	11.3
mercury	13 600	13.6
platinum	21 500	21.5
water	1000	1.0

8 A density bottle has a mass of $50\ g$ when empty. Its mass is $85\ g$ when filled with $25\ cm^3$ of a liquid. What is the density of the liquid?

1.4 Pressure

Pressure is the *force* acting on a unit area.

pressure = $\frac{\text{force}}{\text{area}}$

The units of pressure are newtons per square metre (N/m^2) or pascals (Pa) where $1\ Pa = 1\ N/m^2$.

Pressure and weight

A rectangular block standing on a table exerts a force on the table equal to its weight. The weight of the block does not change whichever way the block stands. The pressure exerted by the block *does* depend on which face of the block is in contact with the table. For example, a block measuring $0.2\ m \times 0.1\ m \times 0.08\ m$ weighs 25 N and stands on a table.
When the face measuring $0.2\ m \times 0.1\ m$ is in contact with the table, the pressure exerted on the table is calculated as follows.

pressure = $\frac{\text{force}}{\text{area}} = \frac{25}{0.2 \times 0.1} = 1250\ N/m^2$

If the face measuring $0.1\ m \times 0.08\ m$ is in contact with the table,

pressure = $\frac{\text{force}}{\text{area}} = \frac{25}{0.1 \times 0.08} = 3125\ N/m^2$

When the face measuring $0.2\ m \times 0.08\ m$ is in contact with the table,

pressure = $\frac{\text{force}}{\text{area}} = \frac{25}{0.2 \times 0.08} = 1562.5\ N/m^2$

Three different pressures may be exerted, but the weight of the block throughout is always 25 N.

Pressure exerted by gases

Gases are tiny fast-moving particles which collide with each other and with the walls of their container. The pressure of the gas is the force exerted by the particles on a unit area of container wall.

If a fixed mass of gas is compressed into a smaller volume, then the number of collisions per unit area increases: in other words, the pressure goes up. This is the basis of Boyle's Law, discussed in more detail on page 56.

Similarly, when the temperature of a fixed mass and volume of gas is increased, the particles move more quickly and strike the container with greater force. Again the pressure goes up.

Atmospheric pressure

The atmosphere is the name given to the thin layer of air that surrounds the Earth. Air has mass and weight, and the weight of the atmosphere exerts a pressure of 100 000 N/m^2 on the surface of the Earth. This pressure is known as atmospheric pressure.

The atmosphere is not uniform: its density decreases with increasing altitude.

The inside of a high-flying jet aircraft has to be pressurized so that the passengers can sit in normal atmospheric pressure.

Atmospheric pressure demonstrations

a) Crushing a can

If the air is sucked from an oil can with a vacuum pump, the can collapses. Before pumping, the pressures inside and outside the can are both atmospheric pressure. When some of the air is removed from the can, the inside pressure falls. The greater pressure outside crushes the can.

b) The Magdeburg hemispheres

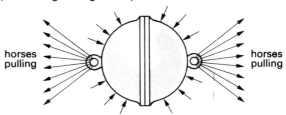

The air was removed from the hemispheres and the horses could not overcome the atmospheric pressure forcing the hemispheres together. As soon as the air was re-admitted, the hemispheres fell apart.

c) The drinking straw

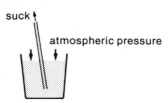

The pressure inside the drinking straw is lowered by sucking. The atmospheric pressure on the surface of the liquid then forces it up the straw and into the mouth.

d) The syringe

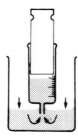

The pressure inside a syringe is lowered by withdrawing the piston. Atmospheric pressure on the surface of the liquid then pushes the liquid up into the syringe.

e) The rubber sucker

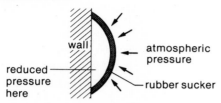

When the sucker is pushed against a flat surface, air is forced out from underneath the sucker. There is less air underneath the sucker and therefore the pressure is reduced. The greater atmospheric pressure outside holds the sucker in place.

Pressure in liquids

The diagram shows a measuring cylinder containing a liquid of density ρ to a depth h. The liquid exerts a pressure on the base of the cylinder. The area of the base is A.

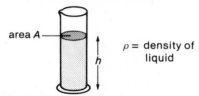

mass of liquid = $A \times h \times \rho$
weight of liquid = $A \times h \times \rho \times g$ (g = 10 N/kg)
But the weight is the force exerted on the base area.

pressure of liquid = $\dfrac{\text{force}}{\text{area}} = \dfrac{A \times h \times \rho \times g}{A} = h\rho g$

Notice that the pressure exerted by the liquid depends on the density of the liquid and its depth, but not on the area over which it acts.

Example

Calculate the pressure exerted by 0.76 m of mercury (density 13 600 kg/m³) in a barometer tube.

pressure of mercury = $h\rho g$
= 0.76 × 13 600 × 10
= 103 360 N/m²

Pressure and depth

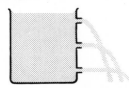

Pressure increases with depth in a liquid. The liquid shoots out farthest from the bottom hole because the pressure is greater there.

Pressure and direction

Pressure at any point in a liquid acts equally in all directions. When the piston in the diagram is pushed, the liquid shoots out equally in all directions.

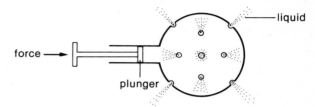

The U-tube manometer

Water in a U-tube settles with the levels in the two halves exactly the same. The water moves until the pressures exerted in each half are equal. We say that 'the water has found its own level'.

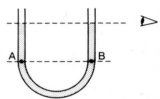

Since the pressure at a point in a liquid depends on the depth of the point below the surface, the pressures at A and B in the diagram are exactly the same. This is true for any pair of points at equal depths. The U-tube can be used to measure the pressure of a gas. It is then known as a U-tube manometer.

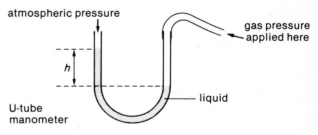

The gas forces the liquid to move to a new equilibrium position so that the pressure of the gas supply is balanced by atmospheric pressure plus the pressure of the column of liquid of height h.

pressure of gas = atmospheric pressure + pressure of h

The manometer tube may contain water, xylene or mercury. Water is used for pressures similar to the domestic gas supply. Mercury is used when higher pressures are to be measured and xylene is used when lower pressures are to be measured.

Measuring atmospheric pressure

Atmospheric pressure is measured using a barometer. There are two main types: the mercury barometer and the aneroid barometer.

The simple mercury barometer is quickly made from a long glass tube, a glass dish and a bottle of mercury. The glass tube should be about one metre long with a diameter of one centimetre. It is sealed off at one end. The tube is filled to the brim with mercury and tapped gently to dislodge any air bubbles. The open end of the tube is covered and the tube is then turned upside down. The covered opening is submerged in more mercury in the dish and finally the cover is removed.

rately measured from a pointer. The top of the mercury column is measured with a vernier.

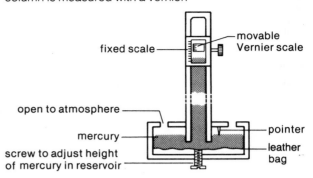

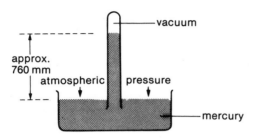

The diagram shows the simple mercury barometer in use. The vacuum exerts no pressure on the mercury in the tube.

The mercury has been forced upwards and is maintained in its position by atmospheric pressure on the mercury in the dish. The height of the mercury (about 0.76 m or 760 mm) is a measure of atmospheric pressure.

Tilting the tube causes the mercury to move up the tube until the height of the column is again 0.76 m (760 mm). The vertical height of the column is only reduced if the space at the top of the column is filled.

The aneroid barometer

The aneroid barometer has a corrugated circular metal box containing air at low pressure. Atmospheric pressure changes cause changes in the shape of the box rather like the collapsing oil can on page 4. The changes are only very small in this case and are enlarged by a system of levers which operates a pointer. The pointer moves over a scale calibrated in units of atmospheric pressure.

The pointer gives an indication of the weather because in general a high atmospheric pressure means fine weather and a low pressure means rain and clouds. A rapid fall in pressure means that there is likely to be a storm.

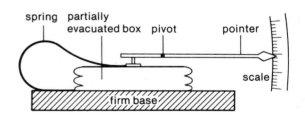

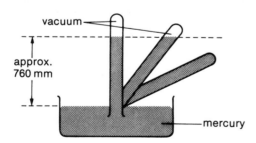

The width of the tube does not affect the height of the column.

The Fortin barometer

The Fortin barometer provides a very accurate means of measuring atmospheric pressure. The reservoir of mercury is contained in a leather bag rather than a dish. The surface level of the mercury can be adjusted and accu-

Exercises

1 Five blocks of polystyrene are placed on the ground. Copy and complete the table below that gives some of the details about each block.

force (N)	area in contact with the ground length (m)	breadth (m)	pressure (N/m²)
72	2	3	
24		2	8
	2	1	48
	1.5	3.5	20
52	3.2		6.5

2 A rectangular block weighs 1.5 N and measures 0.25 m × 0.1 m × 0.3 m. Calculate the pressure on the bench when it stands on
  a) the 0.25 m × 0.1 m face
  b) the 0.1 m × 0.3 m face
  c) the 0.3 m × 0.25 m face

3 The diagram shows a U-tube manometer connected to a gas supply to measure its pressure.

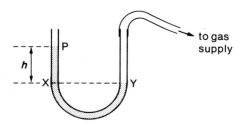

  a) What is the pressure at P?
  b) Is the pressure at Y equal to, more than, or less than the pressure at X?
  c) Is the pressure of the gas supply more than, less than, or equal to atmospheric pressure?
  d) If h = 55 mm and atmospheric pressure is 760 mm of mercury, write down the pressure of the gas supply in terms of millimetres of mercury.

4 The diagram shows three jars containing water, turpentine, and mercury.

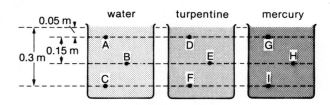

Using the density of the liquids shown in the table below, calculate the pressure at each of points A to I caused by the respective liquid only.

liquid	density (kg/m³)
water	1000
turpentine	800
mercury	13 600

5 The odd-shaped tank contains water (density 1000 kg/m³).

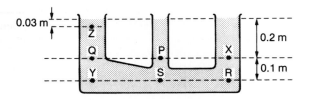

If the pressure at X due to the liquid is 2000 N/m², what is the pressure at the points marked P, Q, R, S, Y, and Z?

1.5 Archimedes' principle

Upthrust

When a solid is immersed in water, it displaces its own volume of water and the water level rises. If the solid object is weighed in air and then weighed in water, it appears to lose weight on immersion. The weight is in newtons.

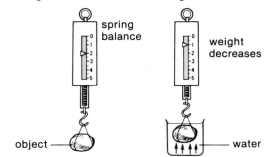

The object has an upward force from the water acting on it. This upward force is known as the *upthrust*. The weight of the object in water is known as the *apparent weight*.

apparent weight = weight in air − upthrust

Archimedes' principle

Archimedes' principle states that
the upthrust experienced by an object immersed in a liquid is equal to the weight of the liquid displaced.

This can be shown to be true with the following experiment.
  a) Weigh a suitable object in air. Weight is measured in newtons.
  b) Weigh the same object in water in a displacement can and measure the volume of water displaced.
  c) Weigh the water displaced.
  d) Check that the apparent loss in weight when the object is weighed in water is equal to the weight of water displaced.

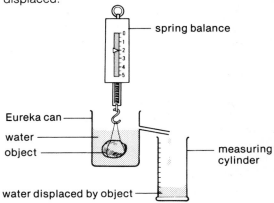

Archimedes' principle also applies to objects immersed in gases.

Sinking and floating

If an object is not hollow but is of uniform density, it will either sink or float when fully immersed in a liquid. Which it does will depend both on the density of the object and on the density of the liquid.

W is the weight of the object
F is the upthrust on the object

When F is greater than W, the object floats.
When F is less than W, the object sinks.

Example

Calculate the upthrust on a block of mass 500 g and volume 250 cm³ when it is immersed in turpentine of density 0.8 g/cm³. Will the block sink or float?

The block displaces a maximum of 250 cm³ when it is totally immersed in the turpentine.

The maximum upthrust that the turpentine can give to the block is the weight of 250 cm³ of turpentine.

Because the density of turpentine is 0.8 g/cm³,
mass of 250 cm³ of turpentine = 250 × 0.8 = 200 g
one kilogram has a weight of 10 N
∴ 200 g weighs 2 N

The turpentine can give a maximum upthrust of 2 N.

The block has a mass of 500 g and weighs 5 N.
5 N is greater than 2 N, so the block sinks.

The hydrometer

A hydrometer measures the relative density of a liquid in which it floats.

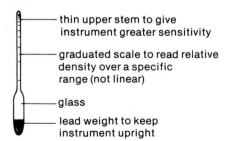

- thin upper stem to give instrument greater sensitivity
- graduated scale to read relative density over a specific range (not linear)
- glass
- lead weight to keep instrument upright

The relative density of the liquid is read off the scale on the stem against the level of liquid. The deeper the hydrometer floats in the liquid, the smaller is the relative density of the liquid. The stem is thin to ensure greater accuracy. The scale on the stem is not linear. The bulb is weighted with lead to keep the hydrometer upright in the liquid.

The hydrometer shown below is used to check the strength of acid in a car battery. A fully charged battery contains acid with a relative density of 1.25 to 1.30.

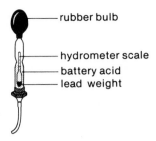

- rubber bulb
- hydrometer scale
- battery acid
- lead weight

When the rubber bulb is squeezed and released, acid is sucked up into the outer glass tube. The miniature hydrometer floats in the acid and registers the relative density of the acid. The acid may then be expelled back into the battery by squeezing the rubber bulb again.

Exercises

Density of water = 1000 kg/m³ or 1 g/cm³
Density of turpentine = 800 kg/m³ or 0.8 g/cm³

1 An object of mass 250 g and volume 750 cm³ is held under water.
  a) What volume of water is displaced by the object?
  b) What is the weight of water displaced?
  c) What is the upthrust acting on the object?
  d) If released, will the object sink or float?
  e) Find the density of the object (i) in g/cm³ (ii) in kg/m³

2 The density of a liquid X is 0.7 g/cm³. An object of mass 500 g and volume 600 cm³ is immersed in water and then in liquid X.
  a) What is the volume of fluid displaced in each case?
  b) Find the mass of water and the mass of X displaced.
  c) What is the upthrust on the object (i) in water and (ii) in X?
  d) Does the object float (i) in water (ii) in X?
  e) Calculate the density of the object.

3 A metal has a density of 8000 kg/m³. What mass of this metal will displace 1000 cm³ of turpentine?

4 5 cm³ of water is displaced by a lump of metal.
  a) Calculate the mass of the lump if the density of the metal is 20 g/cm³.
  b) What volume of turpentine would the lump displace?
  c) Calculate the upthrust on the lump (i) in turpentine and (ii) in water.

5 The weight of an object in a vacuum is recorded as 1 N. If the density of the object is 5 g/cm³, what would be the recorded weight of the object when weighed in an atmosphere of carbon dioxide of density 0.002 g/cm³?

6 Describe how to make a simple hydrometer using a test tube and some lead shot. Make a list of all the things that you would need. What does a hydrometer measure?

7 Draw a sketch of how a hydrometer is used to measure the relative density of turpentine. Now sketch another diagram of the same hydrometer being used to measure the relative density of water.

1.6 Extension of a spring

Stretching

When a weight is loaded onto a suspended spring the spring extends (stretches). The extension of the spring is the increase in the length of the spring compared to its unstretched length.

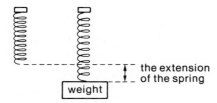

Elasticity

The spring is said to be elastic if it returns to its original length when the stretching load is removed. The spring can be loaded with such a large weight that it is permanently stretched. The maximum weight that can be loaded on the spring so that it just remains elastic is called the elastic limit of the spring.

Load versus extension graph

If the extension of a spring produced by adding increasing weights is plotted against the value of the added weights, a graph similar to the one given below emerges; E represents the elastic limit.

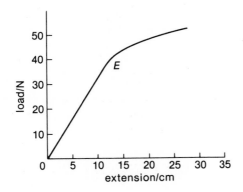

The results for loads less than the elastic limit are summarised as Hooke's Law:
The extension of a spring is proportional to the load applied to it.
This law applies to other elastic materials.

Exercises

1 Draw a load/extension graph for a spring using these data.

load (N)	0.5	1.0	1.5	2.0	2.5	3.0	3.5	4.0
extension (m)	0.02	0.04	0.06	0.08	0.10	0.12	0.14	0.16

Use your graph to answer these questions.
   a) Does the spring obey Hooke's Law?
   b) What load would be needed to produce an extension of 0.05 m?
   c) What load would be needed to produce an extension of 0.125 m?
   d) Find the load on the spring when it has extended 0.0675 m.
   e) What would be the extension for the following loads?
      i) 1.2 N   ii) 3.25 N   iii) 0.25 N   iv) 1.55 N

2 Draw a graph from the table below to show how the total length of a spring changes when it is loaded. Plot a graph of length against load.

length (m)	0.20	0.25	0.30	0.35	0.40	0.45	0.50
load (N)	1.00	2.00	3.00	4.00	5.00	6.00	7.00

Use your graph to answer these questions.
   a) Does the spring obey Hooke's Law?
   b) Find the load needed to produce a spring length of
      i) 0.325 m   ii) 0.42 m
   c) Find the length of the spring when the load on it is
      i) 2.5 N   ii) 6.25 N
   d) How long is the spring when it has no load on it at all?
   e) Find the extension of the spring for a load of
      i) 4.0 N   ii) 1.5 N   iii) 2.4 N

3 Use the data below to plot a load/extension graph for a spring

load (N)	0.0	0.25	0.50	0.75	1.00	1.25	1.50
extension (m)	0.0	0.1	0.2	0.3	0.45	0.75	1.25

Use your graph to answer these questions.
   a) Does this spring obey Hooke's Law all the time?
   b) If your answer to a) is NO, state the range of load over which the law is obeyed.
     If your answer to a) is YES, what extension would you expect the spring to show for a load of 3 N?
   c) Would a similar spring be suitable for use in a spring balance measuring in the range 0–2.00 N?

2 Mechanics

2.1 Moments

The moment of a force

The moment of a force is a measure of its turning effect. It is the product of the force and its distance from the turning point (fulcrum).

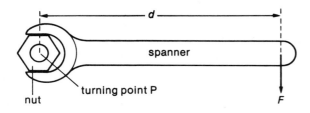

The spanner turns the nut when force F is applied. The nut revolves around the fulcrum, P.
Now,

moment of a force =
force × perpendicular distance from turning point

For the spanner,

the moment of $F = F \times d$

If the force is measured in newtons and the distance in metres, then the moment is measured in newton-metres (N m).

If the force is increased then the moment is also increased. Similarly if the distance is increased the moment is also increased. This latter effect is made use of in a spanner where a long handle increases the moment (turning effect) of the force which is applied.

The principle of moments

If a number of forces on an object is in equilibrium (balanced) then two rules apply:
a) The sum of all the forces in one direction equals the sum of all the forces in the exact opposite direction.
b) The sum of all the moments trying to turn the object clockwise equals the sum of all the moments trying to turn it anticlockwise.
The second rule is known as the principle of moments.

Experimental proof of the principle of moments

Both of the rules of the principle of moments can be verified with the equipment shown below.

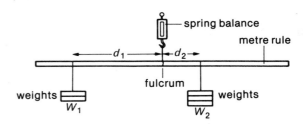

a) Suspend the spring balance and ruler with strong thread from a sturdy support and balance the metre rule before adding the weights.
b) Add weights, W_1 and W_2, attached with more thread and adjust their positions on the ruler until balance is restored.
c) Check the two rules.
  i The reading on the spring balance will be the same as the sum of W_1 and W_2
  ii anticlockwise moment = clockwise moment
  $$W_1 \times d_1 = W_2 \times d_2$$

Exercises

1 State the principle of moments.
2 What is the moment of a force? How do you calculate it?
3 These beams are all balanced. Find the unknown forces or distances marked on the diagrams.

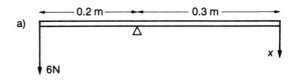

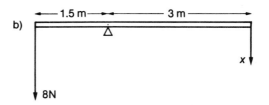

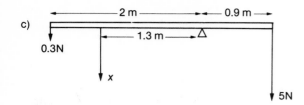

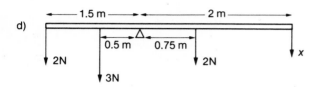

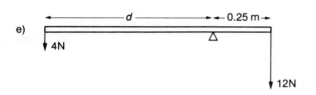

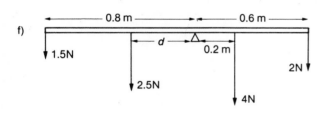

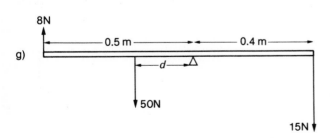

2.2 Centre of gravity

Centre of gravity

An object can be thought of as being made up of lots of particles. Every particle experiences a force due to the attraction of gravity. The sum of these forces is known as the weight of an object. The point through which the weight of an object can be thought to act is known as the centre of gravity (sometimes known as the centre of mass).

If a ruler is supported immediately below its centre of gravity it will balance. Similarly a thin piece of wood or cardboard (a lamina) can be balanced on the tip of a pin if the pin is placed at the centre of gravity.

Finding the centre of gravity of a piece of cardboard

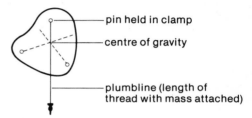

Hang the card and a plumbline from a pin held in a clamp. When the cardboard and plumbline have stopped moving, mark the vertical line given by the plumbline on the card. Then re-hang the card from two other points and mark the vertical lines in the same way. The three lines drawn should then all meet at the centre of gravity.

The weight of a metre rule can be measured using the fact that the weight can be thought to act through the centre of gravity of the rule.

Balance the rule by a thread suspended from a clamp stand. This determines the position of the centre of gravity.

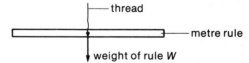

Next, suspend a weight of 1 N by a thread from one end of the rule and adjust the position of the first thread to rebalance the rule.

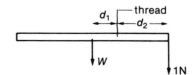

By the principle of moments

$$W \times d_1 = 1 \times d_2 \qquad \therefore W = \frac{d_2}{d_1}$$

4 Describe an experiment to show that the principle of moments is true.

5 You are given a metre ruler, a 100 g mass, a knife edge as a fulcrum, some string and a mass of approx. 100 g labelled X. Describe how you could determine the mass of X accurately using only this apparatus.

Stability

If an object is not easily tilted over, then it is said to be stable. The position of the centre of gravity of an object determines its stability. An object can be in one of three states in relation to its stability. It can be in stable, unstable or neutral equilibrium.

Stable equilibrium

An object in this state falls back to its original position if it is slightly moved and released.

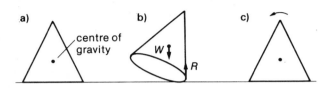

Diagram (b) shows that there is an overall moment produced by the weight *W* and the reaction of the bench *R* to turn the cone back onto its base. A stable object has a wide base compared to its height and a low centre of gravity.

When any stable object is tilted, its centre of gravity moves to a higher position.

Unstable equilibrium

An object in this state does not return to its original position if it is slightly moved and released; it continues to fall.

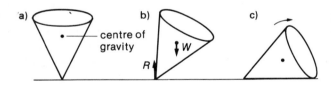

Diagram (b) shows that, in this case, the weight *W* and reaction *R* produce a turning effect which continues the fall of the cone.

When any unstable object is tilted, its centre of gravity moves to a lower position.

Neutral equilibrium

An object in this state stays in its new position if it is slightly moved and released.

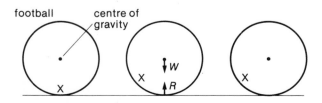

The weight *W* and reaction *R* are always exactly opposite and produce no turning effect and so the football stays in its new position.

Exercises

1. What is meant by the term 'centre of gravity'?
2. Describe how you could find the centre of gravity of an irregularly shaped card.
3. Why does a wise cyclist carry a load in the packs on his bicycle and not in a rucksack on his back?
4. Which vase has the more sensible shape? Explain your answer.

5. XY is a uniform plank. It is shown balanced at F in the diagram. Calculate its weight.

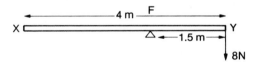

6. Find the weight of the uniform wooden beam shown in the diagram below. The beam is balanced at F.

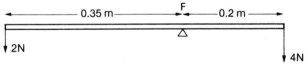

2.3 Vector and scalar addition

Scalar quantities
A scalar quantity has size but no direction. Examples include mass, volume, temperature, time, wavelength and speed. Scalar quantities are added just as any two numbers can be added.

Vector quantities
A vector quantity has size and direction. Examples include velocity, force, acceleration, momentum and weight. When several vectors act together from a point, their combined effect is called their resultant. The resultant is a single vector that can replace those several vectors. For example, two forces of 3 N and 4 N can pull in the same direction, oppose one another or work together at a certain angle.

In a) the resultant is 1 N;

In b) the resultant is 7 N.

The vectors in (c) cannot be added together in such a simple way. We use the parallelogram law.

Parallelogram law
The two vectors to be added are drawn to scale to be the adjacent sides of a parallelogram. The parallelogram is completed and the addition of the two vectors is represented by the diagonal of the parallelogram.

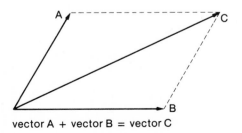

vector A + vector B = vector C

Verification of parallelogram law
In this experiment the vectors chosen are forces.

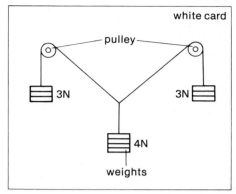

a) Allow the weights and string to settle in an equilibrium position.
b) Mark the position of the strings on the white card.
c) Remove the card and draw the lines clearly with their lengths representing the weights supported.

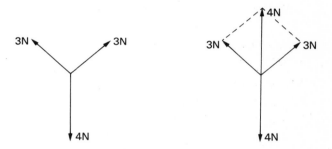

If the arrangement is in equilibrium, then the 4 N weight must be equal and opposite to the resultant of the two 3 N weights. The parallelogram is completed as shown on the right above and, in this case, the arrangement verifies the law.

Resolution
This is the reverse of the addition of two vectors. One vector is resolved into two others having the same effect. The two forces are called the components of the original single force.

The two components are drawn at right angles to each other so that they are easier to calculate.

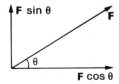

The two components of **F** are **F** cos θ and **F** sin θ

Example

What are the components of a force of 150 N acting at 30° to the horizontal?

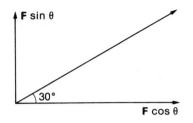

F cos θ = 150 × cos 30 = 150 × 0.867 = 130 N
F sin θ = 150 × sin 30 = 150 × 0.5 = 75 N
It is also possible to solve this problem by a scale drawing of the forces and angle.

Exercises

1. What is the difference between a vector and a scalar quantity? Give two examples of each.
2. State the Parallelogram Law for adding vectors.
3. What is the resultant force acting in these cases?

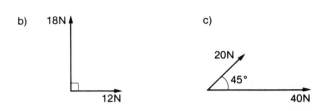

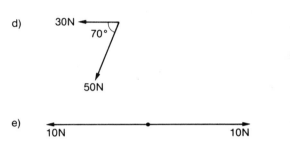

4. Find the resultant of two forces of 10 N and 8 N if the angle between them is 30°.
5. Two vectors of 200 N and 300 N are added. What is their resultant if the angle between them is 120°?
6. What are components? What does 'resolution' mean?
7. Find the components of these forces.

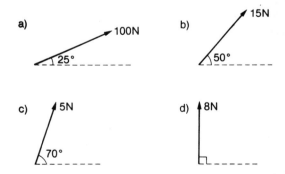

8. A boy flies a kite and attaches a spring balance to read the tension in the string. The balance reads 200 N and the angle between the string and the ground is 60°. What force effectively pulls the boy along the ground and what force pulls him upwards? If the boy weighs 300 N will he be in danger?

2.4 Speed, velocity and acceleration

Speed and velocity

The speed of a moving object is the distance that it moves in one second.

The average speed of the object for a period of motion is found by dividing the total distance moved by the total time of the movement.

$$\text{average speed (m/s)} = \frac{\text{distance (m)}}{\text{time (s)}}$$

The average velocity of an object is also the ratio of distance moved to time taken but in a specified direction. It is also measured in m/s.

Velocity is a vector quantity, i.e. it has direction. Speed, however, is a scalar quantity and does not have a specific direction.

Distance-time graphs

Graphs can be drawn showing the distance covered by a moving object for specified times. They are known as distance-time graphs. A distance-time graph illustrating uniform velocity is given below.

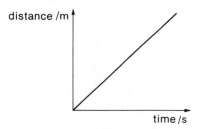

The velocity of the object is the gradient of the line. A distance-time graph of non-uniform velocity may look like the one shown below.

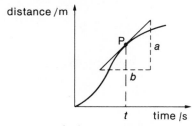

The velocity at instant P is calculated as the gradient of the curve at that instant.

Therefore at time t the velocity $= \dfrac{a}{b}$

Acceleration is the ratio of the change in velocity to the time that the change takes.

$$\text{acceleration (m/s}^2\text{)} = \frac{\text{change in velocity (m/s)}}{\text{time taken (s)}}$$

Acceleration is positive when velocity increases and negative when it decreases.

Velocity-time graphs

Graphs of the velocity of an object against time, called velocity-time graphs, can also be drawn. The diagram below shows a graph of uniform acceleration.

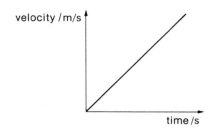

The acceleration of the object is the gradient of the line.

The distance travelled during any given time can be calculated from a velocity-time graph. It is the area between the straight line or curve and the time axis up to the given time.

Example

What is the acceleration of a car and the distance moved in four seconds by the car whose velocity-time graph is as follows?

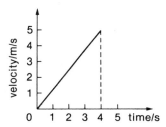

Acceleration = gradient = $\frac{5}{4}$ = 1.25 m/s²

Distance moved after four seconds
= area under curve
= $\frac{1}{2}(4 \times 5)$
= 10 m

Equations of motion

The motion of a uniformly accelerated object can be described by three equations of motion:

a) $\quad v = u + at$
b) $\quad s = ut + \frac{1}{2}at^2$
c) $\quad v^2 = u^2 + 2as$

where u = initial velocity in m/s,
v = velocity after time t in m/s,
a = uniform acceleration of the object in m/s²,
s = distance moved after time t in m.

Example

A ball is accelerated at 5 m/s² from an initial velocity of 2 m/s for a time of five seconds.

What is the final velocity and distance moved?

Final velocity $v = u + at$
$= 2 + (5 \times 5)$
$= 27$ m/s

Distance moved $s = ut + \frac{1}{2}at^2$
$= (2 \times 5) + \frac{1}{2}(5 \times 5^2)$
$= 10 + \frac{1}{2}(125)$
$= 10 + 62.5$
$= 72.5$ m

The distance moved can also be calculated as the area under the velocity-time curve for the acceleration.

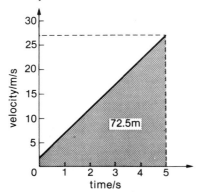

Acceleration due to gravity

An object allowed to fall freely under the influence of Earth's gravity drops with an acceleration g of about 10 m/s².

The value of g can be measured with the apparatus shown in the diagram.

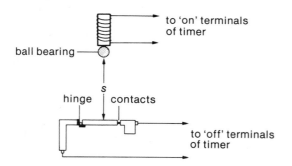

When the electromagnet is switched off, the ball bearing drops. At the same moment the timer is switched on.

The ball drops from rest under the influence of gravity and breaks the lower contacts, switching off the timer. The distance s is accurately measured and g can be calculated from the second equation of motion.

$s = ut + \frac{1}{2}gt^2$
$= 0 + \frac{1}{2}gt^2$
$\therefore g = \frac{2s}{t^2}$

The effect of air resistance

The value of g does not depend on the mass of the dropping object. If a large mass and a small mass are dropped from a height simultaneously they should be expected to hit the ground at the same time.

This was demonstrated on the Moon by one of the Apollo astronauts who dropped a feather and a hammer under the Moon's gravity with the same acceleration.

The same result cannot be achieved out in the open air on the Earth's surface. The resistance of the air to the movement of objects through it has an unequal effect on objects of differing shape and mass. In the case of the hammer and feather, the hammer falls more quickly.

Components of velocity

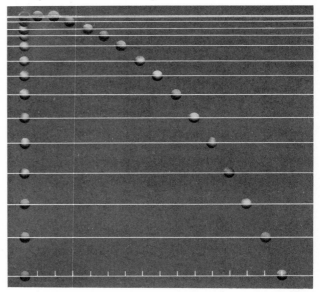

The ball projected to the right moves equal distances sideways in between each flash of light. At the same time both balls move downwards with equal acceleration.

Horizontal and vertical components of a velocity can be considered to be independent.

Tickertape timers

The tickertape timer makes 50 dots per second onto a paper tape that is pulled through the timer. The size of the gap between successive dots depends on the speed of the tape. If the tape moves quickly, the dots are far apart; if it moves slowly, they are close together. When the tape is attached to a moving vehicle, the tape has a record of the motion of the vehicle printed on it.

Analysing tickertape

The complete tape is cut into lengths with an equal number of dots on each length. The lengths represent equal time intervals. The lengths are stuck onto a piece of paper in the same order that they were in on the whole tape.

The following example shows the motion of a tape that increased in speed.

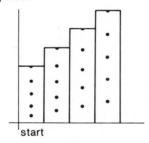

The speed and acceleration of the motion can be found from the tape.

a) The speed during the printing of any length of tape is found by dividing the length of the tape by the time interval that it represents.
b) The acceleration is the difference between any two speeds (as measured in a) divided by the time interval between them.

Example

A, B, C, and D are lengths cut from a whole tape pulled by a moving trolley that passed through the timer. Each length represents a time interval of 0.1 s.

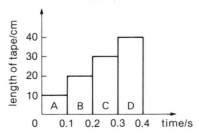

Calculate the speed of the trolley during the production of the lengths A and D and the acceleration of the trolley.

$$\frac{\text{average velocity}}{\text{represented by A}} = \frac{\text{length of A}}{\text{time interval}} = \frac{10 \text{ cm}}{0.1 \text{ s}} = 100 \text{ cm/s}$$

$$\frac{\text{average velocity}}{\text{represented by D}} = \frac{\text{length of D}}{\text{time interval}} = \frac{40 \text{ cm}}{0.1 \text{ s}} = 400 \text{ cm/s}$$

$$\frac{\text{acceleration between}}{\text{A and D}} = \frac{\text{velocity of D} - \text{velocity of A}}{\text{time interval}}$$

$$= \frac{400 \text{ cm/s} - 100 \text{ cm/s}}{0.3 \text{ s}}$$

$$= 1000 \text{ cm/s}^2$$

Exercises

1 Copy and complete the table

distance travelled (m)	time taken (s)	speed (m/s)
10	2	
	5	3
	45	50
24.5	3.2	

2 A boy cycles at 10 m/s for 32.5 s. How far does he travel?

3 Draw a distance-time graph from these data:

distance (m)	10	20	30	40	40	40	50	60	60
time (s)	1	2	3	4	5	6	8	10	12

a) Find the speed during each section of the graph.
b) What was the distance covered in each section of the journey?
c) What distance was covered in the first 8 s?

4 The diagram shows a velocity-time graph for a toy car.

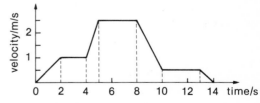

a) How long did the total journey take?
b) What was the maximum speed reached?
c) For how long was the car stationary during the journey?
d) Calculate the acceleration during the first 2 s.
e) How far did it travel during the first 2 s?
f) How far did it travel during the first 5 s?
g) When was the speed increasing fastest?
h) What was the total distance moved?

5 A truck moving at 5 m/s starts to accelerate; 24 s later its speed is 17 m/s. Find its acceleration.

6 A stone falls from rest. If the acceleration due to gravity is 10 m/s^2, how fast will it be travelling after a) 0.75 s, b) 1.25 s, c) 2 s?
Draw a velocity-time graph for the stone and use it to find how far it fell in d) 1 s, e) 1.75 s

7 A car moves with a velocity of 12 m/s for 8 s. It then accelerates at 2 m/s^2 for 4 s and then travels for a further 20 s with a constant velocity. The car then decelerates uniformly to stop in 15 s.
Draw a velocity-time graph for the journey and use it to find:
   a) the total journey time;
   b) the distance travelled in
      i) the first 3 s,
      ii) the first 30 s,
   c) the total distance travelled.

8 The tickertape was cut into lengths that were printed in successive 0.2 s intervals.

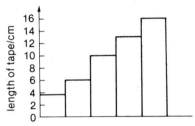

a) Find the average velocity over the first 0.2 s interval.
b) Find the average velocity over the fourth 0.2 s interval.
c) Is the vehicle accelerating? If so, find the average acceleration.
d) What feature of the drawing tells you that the vehicle might be accelerating uniformly?

9 This tickertape was cut into lengths that were printed in successive 0.1 s intervals.

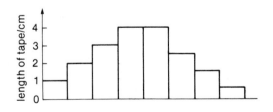

a) Describe the journey of the vehicle.
b) Find the maximum velocity reached.
c) How long did the journey take?
d) Calculate the initial acceleration.
e) Was the deceleration uniform?
f) How far did the vehicle travel?

2.5 Newton's Laws of Motion

Newton's First Law of Motion *Every object continues to be at rest or move with a uniform velocity unless a force acts on it.*

Momentum

The momentum of a moving object is the product of its mass and its velocity.

momentum = mass × velocity

Momentum is measured in N s or kg m/s and it is a vector quantity.

Newton's Second Law of Motion *When a force is applied to a body, its momentum changes. The rate of change of the momentum is proportional to the force.*

$F \alpha \dfrac{mv - mu}{t}$ (α means 'is proportional to')

where F is the force, mv is the final momentum, mu is the initial momentum and t is the period of time for which the force is applied.

From the equation of motion,

acceleration $a = \dfrac{v - u}{t}$

So the Second Law can be written as $F \alpha \, ma$

The newton

The Second Law has led to the definition of the unit of force, the newton.

$F \alpha \, ma$
or, $F = kma$ where k is the constant of proportionality

If we define a newton as that force which produces an acceleration of 1 m/s^2 of a mass of 1 kg, then the constant k becomes one and the equation reduces to: $F = ma$

This remains true for any force, mass or acceleration if F is measured in newtons, m in kilograms and a in m/s^2.

Experimental verification of $F = ma$

The trolley, shown in the diagram, accelerates under the influence of the force.

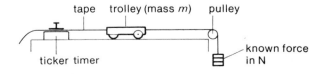

As it accelerates it pulls a tickertape through the timer. The acceleration can be calculated from the tape produced (see p. 17). The mass m of the trolley is found and then the equation $F = ma$ can be tested.

Newton's Third Law of Motion *To every action there will be an equal and opposite reaction.*

Conservation of Momentum

When two objects collide, the total momentum before and after the collision remains constant.

Experimental verification of the conservation of momentum

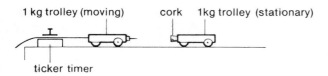

Set the first trolley in motion towards a second, stationary trolley. The first trolley hits the second; they stick together and move off as one. The tickertape will record the velocity of the first trolley before the impact and the velocity of the two trolleys after the impact.

The momentum of the first trolley plus the momentum of the second trolley before collision (which is zero) should be equal to the momentum of the trolleys together after the impact.

Example

A trolley of mass 2 kg moving at 10 m/s hits a second trolley of mass 6 kg moving at 2 m/s in the same direction. The two trolleys stick together and move off at a velocity of 4 m/s.

Momentum before collision

= momentum of 2 kg trolley + momentum of 6 kg trolley
= (2 × 10) + (6 × 2)
= 32 N s

Momentum after collision

= momentum of trolleys together
= (8 × 4)
= 32 N s

This agrees with the conservation of momentum.

Exercises

1 Copy and complete this table

force (N)	mass (kg)	uniform acceleration (m/s^2)
20	4	
	1.5	3
26		2
24	3	
	7	8

2 An object of mass 10 kg has a force of a) 5 N and then b) 0.2 N acting on it. Find the acceleration in each case.

3 A truck of mass 50 kg decelerates at 2 m/s^2. Calculate the force causing this deceleration.

4 A child pulls a toy truck of mass 5 kg with a force of 3 N. What is the acceleration of the truck? How long will it take to reach a speed of 6 m/s if it starts from rest?

5 A force of 8 N retards the motion of a trolley of mass 6 kg which initially moves with a speed of 12 m/s. How long will the trolley take to stop?

6 A bicycle and rider have a combined mass of 90 kg. They accelerate at 3 m/s^2. How big is the accelerating force?

The bicycle slows down with a retardation of 5 m/s^2. What is the size of the braking force?

7 A vehicle of mass 100 kg travels at 15 m/s. The brakes are applied and the speed changes to 9 m/s in 2 s. What is the braking force?

8 A force of 21 N changes the speed of an object from 5 m/s to 14 m/s in 3 s. What is the mass of the object?

9 A trolley has a mass of 2 kg and moves at 3 m/s. It collides with a stationary trolley of the same mass. After the collision both trolleys move together. Find their common velocity.

10 A trolley of mass 5 kg moves at 8 m/s and collides with a stationary trolley. After the collision both trolleys move together with a velocity of 2 m/s. Calculate the mass of the stationary trolley.

11 A boy of mass 40 kg jumps onto a stationary truck of mass 10 kg. The boy and truck move at 3 m/s. What was the speed of the boy when he landed on the truck?

12 A ball A moves at 5 m/s. Another ball B moves in the same direction at 3 m/s; A collides with B and they both continue to move in the same direction at 4 m/s. If the mass of A is 2 kg, what is the mass of B?

2.6 Motion in a curved path

A particle moving in a circle at a constant speed has a continually changing velocity v.

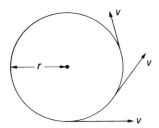

This may seem like a contradiction, but it is not however, because velocity is a vector quantity. It is the direction element of the velocity that is continually changing.

Acceleration

The velocity changes and so there must be an acceleration.

The acceleration a is given by

$$a = \frac{v^2}{r}$$

where v is the velocity of the particle and r is the radius of the circle.

The acceleration is directed towards the centre of the circle.

Centripetal force

From Newton's Second Law,

$$F = ma$$

The force pulling the particle towards the centre of the circle is

$$F = \frac{mv^2}{r}$$

This force is known as centripetal force. It is this force which keeps the movement circular.

Examples of centripetal force

a) A tennis ball tied to a length of string can be whirled around in a circle.
The velocity of the ball changes constantly and the ball accelerates towards the hand. The centripetal force for the circular motion is the tension in the string.

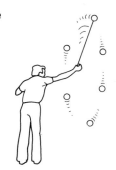

b) The Moon orbits our Earth. The force of gravity on the Moon from the mutual attraction of the Earth and Moon is the centripetal force that maintains the orbit.

Weightlessness

The Earth spins on its axis and we all move with it. A centripetal force must keep us all moving in a circle around the Earth's axis. This force is part of the gravitational force of the Earth. The remainder of the gravitational force provides what we call our weight.

If the Earth started to spin faster, then more of the gravitational force would be needed as centripetal force, and there would be less remaining to provide weight. We would all weigh less.

It is possible to arrange a situation, by flying an aircraft in a particular way for example, where all of the gravitational force is used to provide centripetal force and there is no gravitational force left over for weight. The people inside the aircraft experience weightlessness.

Exercises

1 What is centripetal force?

2 A ball moves in a circle of radius 0.5 m. Its speed is 20 m/s. What is its acceleration towards the centre of the circle?
If the ball has a mass of 2 kg, what is the centripetal force?
If the ball moves in a circle of radius 2 m what are the new values of the acceleration and of the centripetal force?

3 A centripetal force of 75 N acts when a mass of 0.2 kg moves in a circle with a speed of 5 m/s. What is the radius of the circle?

4 A car of mass 1000 kg moves around a circular track at 25 m/s. The track has a radius of 1500 m. Calculate a) the acceleration of the car towards the centre b) the centripetal force acting.

5 What is providing the force necessary to keep each object in the following list circling?
   a) A cyclist going around a corner
   b) A piece of fluff on a record turntable
   c) The Moon going around the Earth
   d) A stone on a string being whirled around someone's head
What would happen in each case if the force acting towards the centre suddenly stopped acting?

6 Astronauts in Earth satellites feel weightless. Explain why this happens.

2.7 Force, work, energy and power

Force

A force is a push or pull.
A force can a) start an object moving,
            b) stop an object moving,
         or c) change the direction of an object's motion.

Work

When the place where a force is acting is moved against the action of the force, work is done. The amount of work done depends on the size of the force to be overcome and how far it is moved.

work done = force × distance moved

Force is measured in newtons, distance in metres and work in joules (J).

Power is the rate at which work is done. A more powerful engine can do more joules of work in a given time than a less powerful one.

$$\text{power} = \frac{\text{work done}}{\text{time taken}}$$

Work is measured in joules, time in seconds and power in watts (W).

Energy

Anything that has energy can do work. Energy is the capacity for doing work.

Energy is also measured in joules.

There are many kinds of energy. Potential energy and kinetic energy are two of them.

Potential energy

An object has potential energy because of its position. It can release this energy by changing position.

If an object has a mass m kilograms, its weight is mg newtons. If the object is lifted upwards from the surface of the Earth by a force to h metres, then the work done against gravity is

work done = force × distance moved
         = $mg \times h$

This work is stored in the object as potential energy, to be released when the object is dropped.

potential energy = mgh (joules)

Kinetic energy

An object has kinetic energy when it is moving. If a stationary object of mass m kilograms is subjected to a force, it will accelerate and reach a velocity v metres/second. The work done on the object to get it to a velocity v is stored in the object as kinetic energy. This is released when the object slows down.

The amount of kinetic energy held by the object depends on the mass of the object and its velocity according to this equation.

kinetic energy = $\frac{1}{2}mv^2$ (joules)

The Law of Conservation of Energy

Energy cannot be destroyed but it can be converted from one form to another.

When a firework rocket is set off, the chemical energy of the rocket fuel is converted into kinetic energy. As the rocket climbs, the fuel is used up and the rate of ascent slows down. The rocket eventually reaches its maximum height. At this point all of the kinetic energy has become potential energy. Then the rocket falls back to the ground. As it falls, its potential energy is converted back into kinetic energy.

Example

What is the potential energy of a 2 kg brick held at a point 5 m from the ground? If it is released and falls to the ground, what is its velocity on impact? ($g = 10$ m/s²)

Potential energy of the brick = mgh
                             = 2 × 10 × 5
                             = 100 J

At impact all of the potential energy (100 J) has become kinetic energy.

Kinetic energy of brick = 100 J
   but $\frac{1}{2}mv^2$ = kinetic energy of brick
   so $\frac{1}{2}mv^2$ = 100 J
   $\frac{1}{2} \times 2 \times v^2$ = 100 J
   v^2 = 100
   v = 10 m/s

Mass and energy

Einstein showed that mass is another form of energy in his famous work on relativity.

He gave an equation for the conversion below.

$$E = mc^2$$

where E is the energy released in joules when a mass of m kilograms is converted
and c is the velocity of light = 3×10^8 m/s.

Exercises

1 How much work does a shopper do lifting a 10 kg bag of groceries onto a table 40 cm from the ground?

2 How big is a force that does 30 J of work when moving a body a distance of 1.5 m?

3 90 J of work are done on a body by a force of 15 N. How far does the body move?

4 A motor is rated at 2 kW. How fast can it lift a load of 50 N through 40 m?

5 A lift raises a load of 30 000 N at a speed of 1.5 m/s. What is the power of the motor?

6 A boy weighing 450 N runs up a flight of 100 steps. Each step is 22.5 cm high and his run is completed in 30 s. How much work does he do against gravity? Find the power he develops.

7 Copy and complete this table.

force (N)	distance moved (m)	work done (J)	time taken (s)	power (W)
5	3		3	
	8	24	1	
1.5		36		9
	100		60	2000
100		100	2	

8 What is the potential energy of a mass of 3 kg held 1.5 m above the ground?

9 What is the change in potential energy when a mass of 2.5 kg is moved from a position 0.5 m above the ground to a position 1.25 m above the ground?

10 The potential energy of a body changes by 850 J when it is moved in the Earth's gravitational field. How much work has been done?

11 What is the kinetic energy of a ball of mass 1 kg moving with a speed of 5 m/s?

12 What is the kinetic energy of a mass of 750 g moving at 4.5 m/s?

13 A ball flying through the air has a kinetic energy of 122.5 J and a mass of 20 kg. How fast is it moving?

14 An object falls from a shelf 3.2 m above the ground. How fast is it moving when it hits the ground?

15 A ball is thrown vertically. It leaves the thrower's hand travelling at 5 m/s. How high will it rise? What will be its speed at the top of its flight and when it returns to the thrower's hand?

2.8 Machines

In a machine, the input force (effort) is different from the output force (load). Machines are often used to move a large load by means of a small effort, but this is only possible if the effort moves through a greater distance than the load.

Mechanical advantage

The ratio of the load to the effort is called the mechanical advantage (M.A.) of the machine.

$$\text{M.A.} = \frac{\text{load}}{\text{effort}}$$

For a lifting machine, this ratio is usually greater than one.

Velocity ratio

The ratio of the distance moved by the effort to the distance moved by the load is called the velocity ratio (V.R.) of the machine.

$$\text{V.R.} = \frac{\text{distance moved by effort}}{\text{distance moved by load}}$$

Again this ratio is usually greater than one in a lifting machine.

When this ratio is less than one, the effort moves a smaller distance than the load. Thus on a bicycle, if the pedals (the effort) move 10 cm, the tyre moves 60 cm and the V.R. is $\frac{1}{6}$ to give a good speed. However, the effort of pedalling is about six times the output force (the load) on the bicycle tyre.

Mechanical efficiency

In a perfect machine the work put in equals the work got out, but machines are usually far from perfect because of friction.

The efficiency of a machine is the ratio of work got out to work put in and is normally stated as a percentage (by multiplying by 100).

work done = force × (distance moved)

$$\therefore \text{efficiency} = \frac{\text{work got out}}{\text{work put in}} \times 100\%$$

$$= \frac{\text{load} \times (\text{distance moved by load})}{\text{effort} \times (\text{distance moved by effort})} \times 100\%$$

$$= \frac{\text{M.A.}}{\text{V.R.}} \times 100\%$$

No machine can possibly give more than 100% efficiency as it would be giving energy for nothing.

Examples of machines

a) The lever

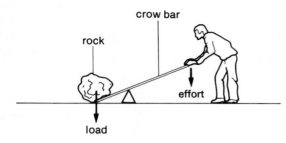

When the clockwise moment produced by the effort is greater than the anticlockwise moment of the load, the rock is raised. The effort made by the hand is smaller than the load of the rock but notice that the hand moves further than the rock.

b) The inclined plane

The inclined plane is a simple machine used to move heavy weights upwards.

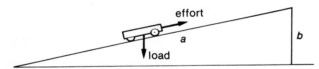

When the load is raised up a vertical height *b*, the effort will have moved a distance *a*. For this inclined plane,

the velocity ratio = $\frac{a}{b}$

the mechanical advantage = $\frac{\text{load}}{\text{effort}}$

c) The pulley system

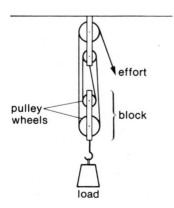

The diagram shows four pulley wheels arranged in two blocks. The load is supported by the string. If the effort is pulled through 100 cm, the load moves up by 25 cm. This pulley therefore has a velocity ratio of 4.

Exercises

1 What is the mechanical advantage of a pulley system when
   a) a load of 60 N is lifted by an effort of 30 N?
   b) a load of 80 N is lifted by an effort of 16 N?

2 A machine has a mechanical advantage of 3. Calculate the effort needed in each case to lift these loads:
   a) 42 N
   b) 25 N
   c) 90 N

3 Find the efficiency of these machines.

mechanical advantage	velocity ratio	efficiency %
4	8	
2	8	
5	5	

4 In an experiment to measure the efficiency of a pulley system, it was found that a load of 60 N was lifted by an effort of 15 N. The distance moved by the effort was 1 m and the distance moved by the load was 0.2 m. Calculate the efficiency of the system.

5 A man lifts a bag of potatoes weighing 250 N into a car boot which is 0.5 m off the ground. How much work does he do on the bag? The farmer has a machine that does the same job and uses up 200 J. How efficient is the machine? What happens to the 'wasted' energy?

6 Two men are loading crated car components into a truck and are using an inclined plane. The plane is 5 m long and is supported on the tail of the truck which is 0.8 m off the ground. They are moving the crates up the plane with the help of a winch. A crate weighing 800 N just moves up the plane when a force of 200 N is applied to it. What is the efficiency of the inclined plane as a machine? How would the efficiency be changed if the crates were placed on rollers?

3 Wave motion and sound

3.1 Waves

Longitudinal waves When an object vibrates, the energy of this vibration travels outwards in waves. For example, the ear picks up the vibrations of the air caused by a sounding bell. The air is called the medium through which the sound travels. The bell causes the particles of air to vibrate in the same direction as the waves travel: sound is a longitudinal wave.

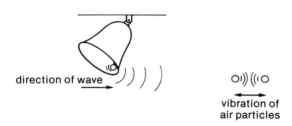

A longitudinal wave of sound can also travel through liquids and solids.

Transverse waves A wave travels across the surface of water. A water wave is a transverse wave because the surface goes up and down in a direction at right angles to the direction of travel of the wave.

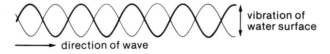

Light and all other electromagnetic waves are also transverse waves.

Displacement-distance graph for water waves. When no wave is travelling, the surface of the water is flat: the water particles are in a position of equilibrium. The height (displacement) of a wave above or below its position of equilibrium is called its amplitude a.

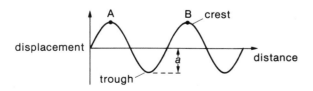

The distance AB between the crest of one wave and the next is called its wavelength λ.

Displacement-time graph The change in displacement of a water particle with time can also be shown on a graph.

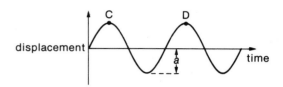

The time CD of one cycle (oscillation) which a water particle takes to return to its first position is shown on the graph. The number of cycles completed in one second is called the frequency f; it is measured in hertz (Hz). One hertz is one cycle per second.

The wave equation The velocity v with which a wave travels through a medium is given by the equation

$$v = f\lambda$$

If the wavelength λ is in metres and the velocity v is in metres per second, the frequency f is in hertz.

Interference

When two waves pass through the same medium at the same instant, they interfere.

If two waves coincide exactly then they interfere constructively.

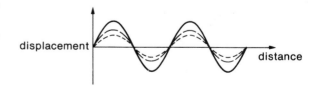

The dark line represents the resultant wave. The two original waves, represented by the broken lines, are said to be in phase.

If two waves that are out of phase coincide, then they interfere destructively.

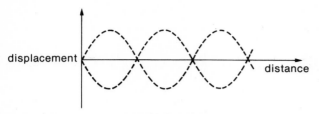

As the two waves are identical in every way, apart from their phases, they cancel each other out and there is no resultant wave.

Two waves that do not have identical wavelengths, amplitudes and a simple phase relationship interfere in a manner in between the two extremes of constructive and destructive interference.

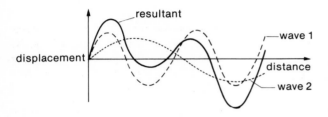

Two wave sources Interference can be shown in the ripple tank between the waves formed by two dippers S_1 and S_2 vibrating close to each other.

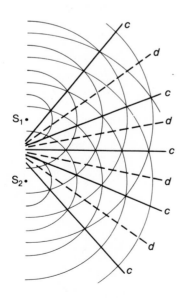

The lines marked c show where there is constructive interference and those marked d show where there is destructive interference.

Young's experiment for measurement of the wavelength of light

When light of one wavelength (monochromatic light) is passed through two narrow slits, an interference pattern is formed on a screen. The pattern consists of dark and bright fringes of light because the two slits act as two identical sources of light S_1 and S_2.

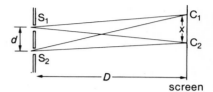

The light from S_2 that forms the bright fringe C_1 has travelled a distance of one wavelength λ further than the light from S_1 that reaches C_1.

d = distance between slits
D = distance from slits to screen
x = distance between two adjacent bright fringes

then $\lambda = \dfrac{dx}{D}$

Young's experiment can be demonstrated with any type of wave, including sound waves.

Diffraction Waves passing through a very narrow slit spread out as though the slit was a point source of the waves; this is called diffraction.

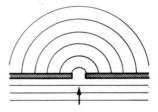

The effect only happens as shown above when the gap is about the same size as the wavelength of the waves passing through it. If the gap is larger than the wavelength, then there is some diffraction (but much less) and only at the ends of the waves as they pass through.

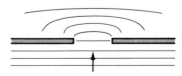

Two slits close together can act as two sources of identical waves which result in an interference pattern.

The diffraction grating The grating consists of a transparent sheet on which are ruled many parallel lines that are close together. There may be as many as 400 lines per mm. The grating acts as a series of very narrow slits which cause diffraction. Some light passes straight through, but the rest is diffracted to form a series of images at different angles to the main path.

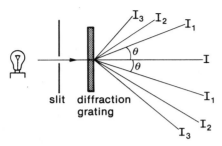

I_1, I_2 and I_3 are the images formed at angles to the main path. I_1 is known as the first order image, I_2 is the second order image, and so on. Note that there are two images of each order, one on either side of the main path and at equal angles from the main path.

The wavelength λ of monochromatic light can be found from the angle θ between the straight-through image I and the first-order diffracted image I_1 as follows.

$$\lambda = d \sin \theta$$

where d is the distance between the parallel lines on the grating. If there are x lines per m on the grating,

$$d = \frac{1}{x} \text{ m}.$$

In general, $n\lambda = d \sin \theta$

where n is the order of the image.

If white light is passed through the grating, the image I is white, but each diffracted image is violet on the side nearer the main path and red on the other side. This is because red light is of longer wavelength and so is diffracted more.

Exercises

1. Give one example each of a) a transverse wave b) a longitudinal wave.
What is the main difference between these two types of waves?

2. Draw a displacement-time graph for a particle in a medium through which a wave is passing.
Use your graph to explain what is meant by the terms
a) oscillation, b) amplitude, c) frequency.

3. Write down the relationship between velocity, wavelength and frequency for a wave motion.

4. Copy this table and fill in the gaps

velocity (m/s)	wavelength (m)	frequency (Hz)
	2	75
300		200
200	4	

5. Two waves pass together through water. One has twice the amplitude of the other and they both have the same wavelength of 4 cm. The smaller amplitude is 2 cm. On the same axes draw a graph of the profiles of the water surface due to the passage of each wave. Show also the resultant wave profile.

6. Explain what is meant by the terms interference and diffraction in connection with wave motion. What conditions are necessary for interference and diffraction to be observed?

7. Wavelength may be measured using Young's Double Slit Experiment.
   a) Explain
      i) why two slits are needed   ii) how wavelength may be found
   b) What happens to the spacing between the fringes if:
      i) the slits are moved further away from the screen   ii) the slits are placed further apart?

8. In a Young's Double Slit Experiment to measure wavelength, the slits were 0.3 mm apart and the screen was 2 m from them. The fringe separation was measured as 0.4 cm. What was the wavelength of the source?

9. Using light of wavelength 5.9×10^{-7} m (0.000 000 59 m), the fringe separation in Young's Experiment was found to be 0.5 cm. If the slits were 0.3 mm apart, how far was the screen from the slits?

10. A diffraction grating is used to measure wavelength.
    a) If there are 600 000 lines ruled per metre on the grating, what is the spacing between the lines?
    b) If the angle moved through from the straight-through position to find the first order image is 20°, what is the wavelength of the source?

3.2 Sound

We hear a sound when something vibrates in air. The vibrations of the sounding object travel through the air to our ears. Ears are able to transform the vibrations received into signals to the brain.

Tuning fork

The prongs of a tuning fork vibrate when struck. They vibrate with a frequency fixed by the size and material of the fork.

The diagram below shows how the movements of one prong of the fork send out a sound wave.

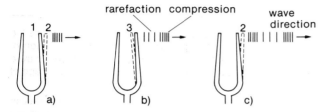

When the prong moves out to position 2, it compresses the air in front of it. As it moves back to position 3, it leaves a volume of rarefied air. When it moves out to position 2 for a second time, it creates another compression, and so on.

The sound wave moves out from the tuning fork as a series of compressions and rarefactions. The wave is passed on by the air molecules. The air molecules vibrate about their equilibrium position.

The vibration of the air molecules is in the same direction as the movement of the wave. This sound is a longitudinal wave motion.

The vibrations of a tuning fork can be felt if it is struck and held lightly against the cheek.

Sound and vacuum

Sound can only travel through a medium. It cannot travel through a vacuum.

If a ringing bell is placed in a bell jar and the air inside the jar is pumped out, the ringing becomes fainter and fainter, and in the end cannot be heard at all.

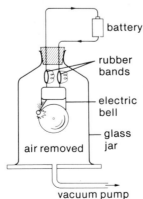

Echoes

Echoes are the reflections of sound waves. It is possible to measure the velocity of sound by timing the return of an echo from a distant building. The velocity is calculated by dividing the total distance that the sound wave moves by the time taken.

Echo sounding is used by seamen to determine the depth of the water that they are sailing on. A high frequency sound wave (an ultrasonic wave) is sent out from the ship. This is reflected back from the sea bed to the ship.

If the velocity of sound in water is 1500 m/s and the signal is received back at the ship in time t, the depth d of the water at that point is given by

$$d = \frac{1500t}{2}.$$

Echo sounding can also be used by fishing boats to find shoals of fish.

Unwanted echoes need to be eliminated in recording studios and concert halls. Echoes can be reduced by covering the walls, ceilings and floors with thick sound-proofing tiles, curtains or carpets.

Resonance

All vibrating systems have a natural frequency of vibration. If the system is made to vibrate at its natural frequency, the amplitude of the vibration quickly builds to a maximum. This effect is known as resonance.

For example, a singer can shatter a glass by singing a certain note. The frequency of the note is the same as the natural frequency of the glass. The singer sings and the glass resonates. The amplitude of the vibrations increases and the glass shatters.

Stationary waves

Longitudinal waves can be made to move down a slinky spring attached to a wall by giving a short sharp push and pull to the end of the spring.

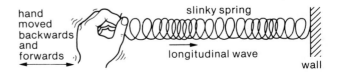

The wave reflects from the wall and moves back down the slinky.

If several waves are sent out in a continuous pattern, the outgoing waves interfere with the reflections. By altering the frequency, it is possible to create the illusion that the compressions and rarefactions are standing still.

This kind of apparently steady vibration pattern is called a stationary wave.

Resonance tube

A vibrating tuning fork is held over the mouth of the tube as the water slowly dribbles out. There comes a point when a loud note is heard coming from the tube. At this point the air in the tube is resonating. Sound waves travel down the tube, reflect off the water and go back up the tube exactly in step with the vibrations of the tuning fork. A stationary wave is set up in the air.

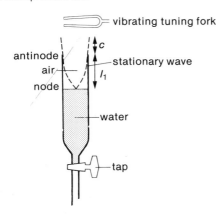

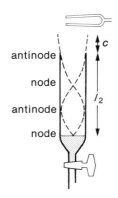

The air at the mouth of the tube is free to vibrate and it does so to its maximum amplitude. The position of maximum amplitude in a stationary wave is known as an antinode. The air in contact with the water does not move. This position is the opposite of an antinode and is known as a node.

The diagram above gives a transverse representation of a longitudinal wave. The node to antinode distance is one quarter of the wavelength. The air in a tube vibrating in this way is said to be vibrating at its fundamental frequency.

The velocity of sound in air can be measured using a resonance tube.

For a resonance tube resonating at the fundamental frequency (see diagram above) the wavelength λ is given by

$$\tfrac{1}{4}\lambda = l_1 + c \quad \ldots [1]$$

where l_1 is the length of the tube and c is an end correction. The end correction allows for the fact that the antinode does not coincide exactly with the end of the tube.

A second position of resonance for the same tuning fork can be found and measured.

In this case $(l_2 + c)$ represents $\tfrac{3}{4}$ of λ.

So, $\quad \tfrac{3}{4}\lambda = l_2 + c \quad \ldots [2]$

Subtracting [1] from [2] eliminates c (which is impossible to measure),

$$l_2 - l_1 = \frac{\lambda}{2}$$

So, $\quad 2(l_2 - l_1) = \lambda \quad \ldots [3]$

If the frequency of the fork f is known, the velocity of sound v can be calculated from

$$v = f \times \lambda$$

Open pipes resonate with an antinode at each end and nodes in the middle. The diagram shows an open pipe vibrating at its fundamental frequency.

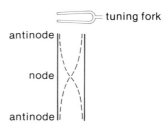

Exercises

1. What kind of waves are sound waves?
2. How can you show that a sounding tuning fork is vibrating?
3. How can you show that sound cannot be transmitted through a vacuum? Sketch the apparatus you would use.
4. What is an echo? A boy stands 250 m from a cliff. He finds it takes 1.5 s for the echo of a clap to return to him. What is the velocity of sound in air?
5. What is resonance? Give an example.
6. If the node-to-antinode distance for a stationary wave is 0.5 m, what is the wavelength?
7. A tuning fork, with frequency 256 Hz, is struck and held over a resonance tube. Resonant positions are found when the length of the air column is 0.31 m and 0.96 m. Calculate the velocity of sound in air and the size of the end correction.

3.3 Vibrating strings

Stationary waves are produced in a taut wire when it is vibrated. A stationary wave is formed when one wave progresses down the string and is reflected back to interfere with and reinforce the next progressive wave.

The sonometer can be used to demonstrate stationary waves.

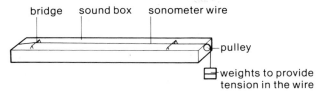

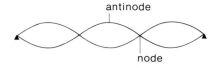

Bow or pluck the wire of the sonometer whilst touching it lightly anywhere along its length. The wire vibrates as shown above. Places where the wire appears not to move are called nodes and places where the wire moves most are called antinodes.

Fundamental frequency

When a wire vibrates (like the one shown below) with an antinode in the middle, we say that it is vibrating at its fundamental frequency.

The value of a fundamental frequency depends on the length of the wire, the tension in the wire, and the mass per unit length of the wire.
a) Increasing the tension in the wire increases the frequency.
b) Increasing the length of the wire lowers the frequency.
c) A string with a high mass per unit length gives a lower frequency than a string with a low mass per unit length.

Stringed instruments, like the violin or guitar, are basically sonometers. The strings are made to vibrate by bowing or plucking and the notes are amplified by the air in the instrument.

Notes of the same frequency played on different instruments sound slightly different because of the variation in the number of overtones played. An overtone is a note of higher frequency which is a multiple of the fundamental frequency.

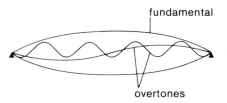

Musical notes

There are three important properties of musical notes: pitch; loudness; quality.
a) Pitch is the musicians' term for frequency. A high note has a high frequency and a low note has a low frequency.

For example: middle C has a frequency of 256 Hz,
upper C has a frequency of 512 Hz.

If two notes are an octave apart, then one has a frequency exactly twice that of the other. Middle and upper C are an octave apart.
b) The loudness of a note is a measure of the amplitude of the vibration producing it.
c) The quality of a note depends on the pitch and loudness of the overtones that accompany the fundamental frequency. So middle C played on a trumpet has a different sound to middle C played on a piano. The fundamental frequency (middle C) is the same for both instruments but the overtones that each instrument provides give the quality of the sound.

Exercises

1 A guitarist plays his instrument. Which of the strings does he pluck for the low notes? What are the frets for? How does he get the highest notes of all? How does he increase the tension in each string and why would he want to do that?

2 What is a sonometer? How is the tension in the wire produced?

3 What is an overtone? A trumpet and a trombone both play middle C, yet they sound slightly different. Explain why that is.

4 A sonometer wire 0.90 m long is plucked in the middle and it vibrates at its fundamental frequency. What is the wavelength of the note? Would you expect to find a node or an antinode in the middle of the wire? If the velocity of sound in air is 330 m/s, calculate the frequency of the note. (Hint: use $v = f\lambda$.)

3.4 The electromagnetic spectrum

Visible light is a small part of a very large family of waves of energy known as the electromagnetic spectrum.

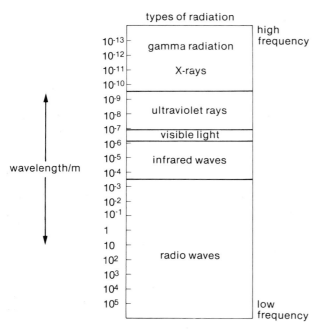

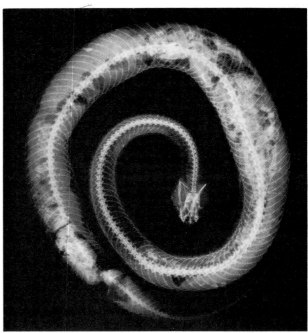

X-rays have many properties: they can penetrate matter, finding use in the detection of broken bones; they can ionise gases; they are diffracted by crystals; they can cause fluorescence and they give the photoelectric effect.

All of the radiations are transverse waves and can travel through a vacuum and move with the same velocity in air, 3×10^8 m/s.

The lines on the diagram separating the different radiations are not meant to represent a rigid boundary. Radiations merge one into the other just like the colours of the visible spectrum (red, orange, yellow etc).

γ-radiation is produced when there are changes in the energy levels in the nuclei of radioactive atoms (see p. 90). They have similar properties to X-rays discussed below; indeed they are very difficult to distinguish from X-rays and there is considerable overlap between the two regions. γ-radiation may be detected with a Geiger counter because of the ionisation of the gas within the Geiger tube, or by photography, or with a cloud chamber.

The γ-rays passing through the cloud chamber collide with atoms in the chamber, ejecting one or more electrons. The electrons produce the tracks by condensing the supersaturated vapour in the chamber.

X-rays are given off when fast moving electrons smash into a tungsten target. They come from energy changes in the electrons most closely held to the nuclei of the tungsten atoms.

Ultraviolet radiation comes from high energy changes in the electron structure of atoms. It is normally produced in the laboratory by passing an electrical discharge through a glass bulb containing mercury vapour.

Ultraviolet radiation may be detected photographically or by fluorescence. The 'mystery' ingredient included in washing powders is often a chemical which fluoresces in ordinary sunlight. Ultraviolet lamps are occasionally used in discos. They appear to be purple but the ultraviolet output has no colour. The radiation picks out white clothing and teeth to give an interesting effect. Unfortunately, it also highlights dandruff!

Ultraviolet can be very harmful. Indeed our atmosphere protects us by absorbing most of the ultraviolet output from the Sun. The small amount that does get through causes all of those lovely summer tans.

Visible radiation is the name given to light. The spectrum runs from violet at about 4×10^{-7} m wavelength to red at about 7×10^{-7} m wavelength. The light is given out when the outer electrons surrounding atoms are rearranged and this happens in very hot solids, in flames or in gas discharge tubes. Visible light is detected by the eye, with a photocell or on the film in a camera. It can cause some other chemical actions to occur.

Infrared radiation This radiation is from the opposite side of the visible light spectrum to ultraviolet and it is consequently less energetic. It comes from low energy changes in electron structure. Everything gives off infrared to a greater or lesser extent but in the laboratory an object which has been heated is used, e.g., gas in a Bunsen flame.

Infrared can be detected photographically and it can be used to take photographs at night or through haze.

It can also be detected with a thermopile and sensitive galvanometer.

The thermopile consists of a lot of thermocouple junctions, each one a join between two different metals or alloys. When infrared radiation reaches the thermopile, a tiny current of electricity is produced and is registered by the galvanometer.

Radio waves are low energy waves produced when electrons are made to oscillate in special electronic devices and they may be detected with a suitable aerial and receiver circuit.

Radio waves, particularly those with longer wavelengths, are used in communications for radio and television broadcasts.

Short wavelength radio waves find application in microwave ovens, in the analysis of atomic structure and in radar.

Exercises

1 Rearrange the following types of radiation in order of increasing wavelength: infrared, ultraviolet, gamma rays, radio waves, X-rays, visible light.

2 Give one method which can detect each of the following: X-rays, ultraviolet light, visible light.

3 Which radiations from the electromagnetic structure are safe to man and which are dangerous?

4 How may X-rays, ultraviolet radiation and radio waves be put to practical use?

5 Why do people who work with X-rays and other radiation wear film badges?

6 All radiations in the spectrum move at 3×10^8 m/s in air. How long do they take to come from the Sun which is about 1.54×10^{11} m away?

7 All the radiations in the electromagnetic spectrum move at the same velocity and they are related by the equation:
velocity = frequency × wavelength.
Use this information to compare the frequencies of infrared and ultraviolet radiation.

4 Light

4.1 Light travels in straight lines

Light energy A beam of light is a stream of light energy. A light ray is the path taken by light. An object can only be seen when light from it enters an eye.

Luminous objects give out their own light and can be seen in the dark. The Sun, other stars and working electric light bulbs are all luminous.

Non-luminous objects can only be seen when light, originally from a luminous object, is reflected from them. A book, a chair and the Moon are all non-luminous.

Light travels in straight lines Shine a torch on a dark night and the beam travels straight out into the darkness. The fact that you cannot see around a corner is further evidence for light travelling in straight lines.

Shadows are another consequence of the straight-line movement of light. The nature of a shadow depends upon the size of the light source casting it.

a) A very small light, known as a point source, casts a very dark shadow of an object. The shadow has a crisp outline and is called the *umbra*:

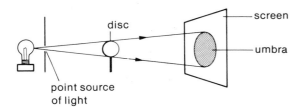

No light from the light source reaches the umbra. Moving the disc towards the light source makes the umbra bigger; moving the disc towards the screen makes the umbra smaller.

b) A large light, known as an extended source, casts a shadow with a much less well-defined edge:

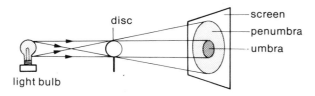

Light from the side of the bulb facing the disc contributes to the formation of the shadow but the diagram is made clearer if the light from two places only is considered. The dark umbra is surrounded by a blurred and grey area called the *penumbra*. No light reaches the umbra, as before, but some light reaches the penumbra.

Eclipses are shadows cast on a grand scale. When the Sun is eclipsed, the Moon passes between the Sun and the Earth and a shadow of the Moon is cast on the Earth:

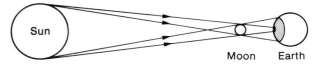

The Sun is an extended light source, so the shadow cast on the Earth has both umbra and penumbra. People in the penumbra see a partial eclipse; people in the umbra see a total eclipse:

Never look directly at the Sun or through any equipment: the eye is *easily* damaged in this way, even during an eclipse.

An eclipse of the Moon occurs when a shadow of the Earth is cast on the Moon.

The pinhole camera

The camera is a light-proof box with a pinhole in one end;

the opposite end is covered with a photographic film or a translucent screen of greaseproof paper or similar material:

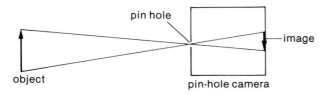

Light travels out in every direction from all points on the object and some of it passes through the pinhole to form the image on the screen. The paths of only two rays are given on the diagram for clarity; the rays cross over at the pinhole and the image is consequently inverted (upside down).

A pinhole camera with two pinholes has two images and one with three pinholes has three images, and so on.

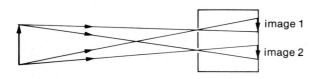

If the pinhole camera is moved nearer the object, the image will be bigger; if it is moved further from the object, the image will be smaller.

When the pinhole is enlarged, the image becomes brighter but more blurred. A blurred image can be focussed with a convex lens held just in front of the enlarged pinhole.

Exercises

1 Explain how a shadow is formed. Use a diagram to illustrate your answer.

2 A girl puts her hand in front of a lamp and casts a shadow on the wall. Describe what happens to the size of the shadow if:
    a) the hand is moved towards the wall
    b) the hand is kept still and the lamp is moved away from the wall
Use diagrams to explain your answers.

3 In the diagram a lamp is shown casting the shadow of the shape of a cardboard star on a screen. Light reaches the shape through a small hole in a piece of card.

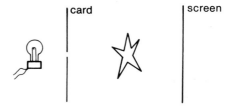

What would be the effect on the shadow of removing the card?

4 Draw a diagram to show how a pinhole camera forms an image.

5 How many images would you expect when using a pinhole camera with twenty pinholes in it? How many of the images would be inverted?

6 State what happens to the sharpness and brightness of a pinhole camera image if the pinhole is made bigger.

7 Draw a ray diagram of a solar eclipse. Indicate on your diagram where you would have to be to see a total eclipse.

8 Draw a ray diagram for an eclipse of the Moon.

9 Describe what you think it feels like to experience a total eclipse of the Sun. Are you likely to experience one?

4.2 Reflection

The laws of reflection

The following diagram shows a ray of light reflecting from a plane (flat) mirror:

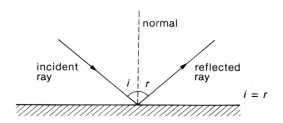

The normal is a line drawn at right angles to the mirror surface.

Here are the two laws of reflection obeyed by light every time it is reflected:
1) The incident ray, the reflected ray and the normal at the point of incidence all lie in the same plane.
2) The angle of incidence is equal to the angle of reflection.

The angle of incidence i is the angle between the incident ray and the normal. The angle of reflection r is the angle between the reflected ray and the normal.

Showing that $i = r$

a) Shine a ray of light across a piece of paper so that it reflects from a plane mirror.
b) Use a pencil to mark in the path of the incident and reflected rays and the position of the mirror.
c) Add the normal at the point where the ray is reflected and measure the angle of incidence and angle of reflection. They should be equal.

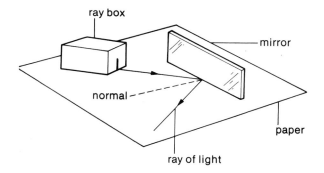

The properties of an image formed by a plane mirror

a) It is the same size as the object.
b) It is the same distance behind the mirror as the object is in front.
c) It is laterally inverted (turned around sideways); see the photograph below.
d) It is virtual, i.e. it cannot be seen on a screen (compare with a concave mirror).

Formation of an image in a plane mirror

Rays of light leave the object and reflect from the mirror. They appear to have come from the image.

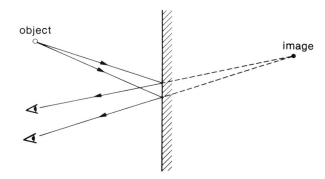

Uses of plane mirrors

a) In the home, e.g. mirrors to look into while combing your hair or mirror tiles used on walls to make a room look bigger.
b) The periscope is used to see over the heads of a crowd. It consists of a tube containing two mirrors arranged so:

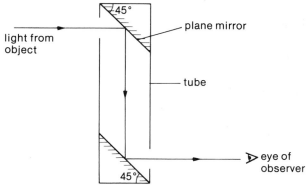

The mirrors are parallel to each other and at 45° to the ends of the tube so that light is reflected down the tube and then again into the eyes of the person using it.

c) In the laboratory when taking readings using a pointer on a scale. The mirror is placed under the pointer and the reading taken at the position where the pointer and its image in the mirror appear as one. The person taking the reading can be sure then that the pointer, its image and the scale are all in line vertically and that the reading is correct.

d) The kaleidoscope is a toy consisting of a tube containing two mirrors, at 60° to one another, and some coloured pieces of plastic. The mirrors produce five images of the coloured objects and, by shaking, the viewer can see an infinite variety of pretty patterns.

Convex mirrors

A convex mirror is a curved mirror having the silvering on the inside of the curve:

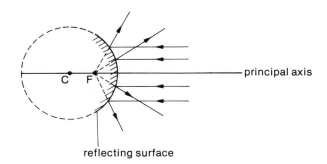

C is the centre of curvature, the centre of the sphere of which the mirror is part. Rays of light obey the laws of reflection when reflecting from a convex mirror. Rays of light parallel to the principal axis appear to be diverging (spreading out) from the focus F of the mirror. The focus is exactly half-way between C and the mirror. The distance from F to the mirror is called the focal length of the mirror.

An image in a convex mirror is always vertical, upright and smaller than the object.

Uses for the convex mirror

a) In supermarkets to deter shoplifters.
b) In buses so that the driver can see passengers coming downstairs.
c) As a driving mirror. A driver can see more with a convex mirror than he can with a plane mirror of the same size.

Concave mirrors

A concave mirror has the silvering on the outside of its curve:

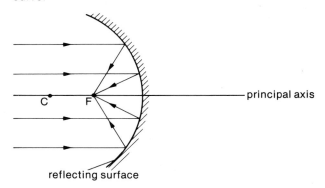

Rays of light, parallel to the principal axis, strike the mirror, converge (come closer) and pass through the focus F. As before, each reflection obeys the laws of reflection and F is half-way between the centre of curvature C and the mirror.

The image produced by a concave mirror depends upon the position of the object in relation to C and F:

a) An object placed beyond C will have an image which is real, upside down and smaller than the object.
b) An object at C will have an image which is real, upside down and exactly the same size as the object.
c) An object between C and F will have an image which is real, upside down and larger than the object.
d) An object between F and the mirror will have an image which is virtual, upright and larger than the object.

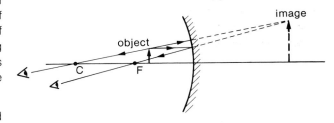

Uses for the concave mirror

a) **As a magnifier**
   The concave mirror is useful when the object is nearer the mirror than the focus and the image is magnified. The dentist's mirror and the shaving mirror are examples of this.

b) To produce a parallel beam of light

If a lamp is placed at the focus of a concave mirror, light rays from the lamp are reflected so that a parallel-sided beam of light is produced. This arrangement is used in car headlamps and pocket torches:

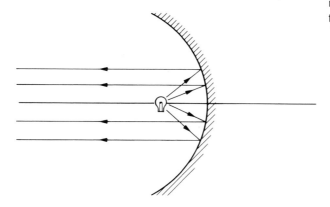

4.3 Refraction

Refraction occurs when light passes from one medium (e.g. air) into another (e.g. water).

If the light passes from a less dense to a more dense medium (e.g. from air to water), then it will be refracted towards the normal.

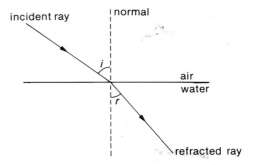

If the light passes from a dense to a less dense medium (e.g. from glass to air), then it will be refracted away from the normal.

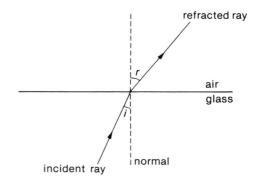

Refractive index

For any example of refraction: the angle between the incident ray and the normal is known as the angle of incidence i; the angle between the refracted ray and the normal is known as the angle of refraction r.

There is a special relationship between i and r for any pair of media known as Snell's Law; it says:

$$\frac{\sin i}{\sin r} = \text{a constant number}$$

The constant is the refractive index for the two media.

The refractive index for air to glass is 1.53 and it has the same value for any pairs of values for i and r for that particular sample of glass.

Exercises

1 Using a protractor, draw an accurate diagram to show the path of a ray of light after reflection at a plane mirror if the angle of incidence is 30°. Label the following on the diagram:
 a) the incident ray
 b) the reflected ray
 c) the mirror
 d) the normal to the mirror
 e) the angle of incidence
 f) the angle of reflection

2 Draw a diagram to show how a plane mirror forms an image of a point object.

3 Draw an accurate diagram to show how a plane mirror forms an image of an object 4 cm long placed 6 cm in front of it.

4 What is lateral inversion? Write your name down as you would see it in a plane mirror. Can you think of a use of lateral inversion?

5 How many mirrors are there in a periscope? How are they arranged? Draw a diagram to show the path of a ray of light through the instrument.

6 What is the difference between a convex mirror and a concave mirror? Which one converges parallel light rays?

7 Where do you put the object in relation to a concave mirror in order to produce a magnified image which is a) real, b) virtual?

8 Which curved mirror would you use for a solar cooker? Where would you put the food?

9 Draw a convex mirror which has a focal length of 4 cm. Mark in the centre of curvature.

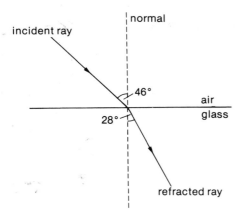

The angle of incidence *i* in this example is 46°.
The angle of refraction *r* is 28°.

$$\frac{\sin i}{\sin r} = \frac{\sin 46°}{\sin 28°} = \frac{0.7193}{0.4695} = 1.53$$

Refraction and the velocity of light

The velocity of light in any one medium is a constant. The velocity of light in a dense medium is lower than it is in a less dense medium.
The actual value of the velocity is related to the refractive index of the medium, for example:

refractive index for air to glass = $\dfrac{\text{velocity of light in air}}{\text{velocity of light in glass}}$

When light enters glass, its velocity and wavelength both decrease but its frequency remains the same.

Refraction and the reversibility of light

A ray of light passing through a parallel-sided glass block takes a path something like this:

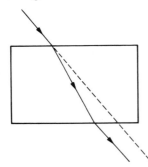

The path of the ray as it emerges from the block is parallel to, but displaced sideways from, the original path.

If the source of the light shown above is turned round and placed on the opposite side of the block so that it shines along the path of the emerging ray, the light will travel back through the block and out again exactly along its original path.

Examples of refraction

a) Apparent decrease in depth of a swimming pool

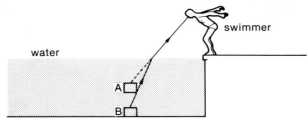

The swimmer about to dive in sees the brick at A. It is really at B: light from the brick is refracted at the water surface into the swimmer's eye. Because light normally travels in straight lines, the swimmer's brain assumes that the brick is at A.

b) The 'bent' stick

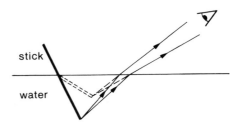

The stick appears to be bent because light from the part of the stick under the water is refracted as it leaves the water surface. The submerged parts of the stick appear to be at a shallower depth than they really are.

The critical angle and total internal reflection

When light leaves a dense medium to enter a less dense medium, it is refracted away from the normal. As angle *i* is increased, there comes a point when angle *r* is 90°:

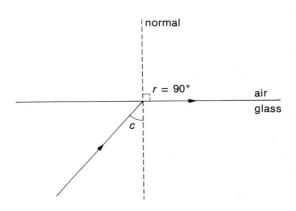

When this happens, the angle of incidence has a special value called the *critical angle c*. The critical angle for glass is about 42°.

The refractive index of a medium can be calculated from its critical angle. If the path taken by the light ray shown above was reversed, the incident angle would be 90° and the angle of refraction would be the critical angle.

Applying Snell's Law:

$$\frac{\sin 90°}{\sin c} = \text{refractive index}$$

As $\sin 90° = 1$,

$$\frac{1}{\sin c} = \text{refractive index of the medium}$$

If the angle of incidence is increased beyond the critical angle, the ray of light undergoes total internal reflection.

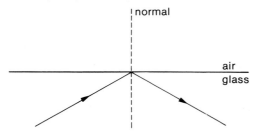

Applications of total internal reflection

a) A periscope with prisms

The periscope contains two right-angled isosceles prisms:

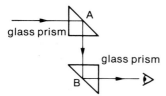

Light is totally internally reflected at A and B. This kind of periscope is used in submarines; the reflecting surfaces are not exposed to tarnishing by air. The same principle is used in prismatic binoculars.

b) Erecting prism

In the right-angled prism below, light rays enter the face AB. Total internal reflection occurs at BC and the light beam leaves the prism through AC with the image turned upside down. The prism may be used to correct the images from a projector lens.

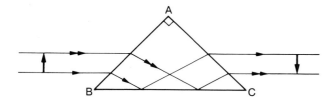

A five-sided prism is used in the view-finder of a single lens reflex camera to correct for the inversion in both the lens and the mirror, as shown below.

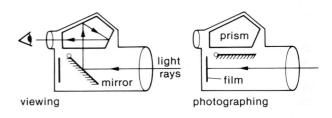

Exercises

1 Draw a diagram to show what is meant by refraction. Draw separate diagrams to show the path of a ray as it passes:
 a) from air to water
 b) from glass to air
 c) from water to glass
(refractive index of water = 1.33; refractive index of glass = 1.50)

2 Calculate the refractive index of perspex from the information below:

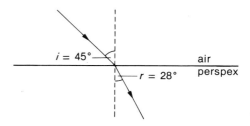

3 Calculate the angle of refraction for a ray of angle of incidence 37° entering a block of refractive index 1.66.

4 What is meant by the term refractive index? If the velocity of light in air is 3×10^8 m/s and the velocity of light in water is 2.3×10^8 m/s, find the refractive index for light passing from air into water.

5 When you look into a swimming pool a stone at the bottom may seem closer to you than it really is. Explain this. Under what circumstances would the stone at the bottom of the pool appear to be at its true distance from you?

6 Explain what is meant by the terms:
a) critical angle
b) total internal reflection
Under what circumstances can total internal reflection occur?

7 Calculate the critical angle for a material of refractive index 1.33.

8 What is the refractive index of a material that has a critical angle of 39°?

9 Explain how a right-angled prism can be used to turn a beam of light through 90°. Give one use of total internal reflection.

10 Why is a periscope made from glass prisms preferred to one made with plane mirrors for use at sea? Draw a diagram to show how the prisms would be arranged and the path of a ray of light through the instrument.

4.4 Lenses

Converging and diverging lenses

Lenses are pieces of glass or plastic having one or more faces curved. When light passes through a lens it is refracted at both faces according to Snell's Law, and this alters the path of the light by an amount which depends on the curvature of the surface.

Lenses that are thicker in the middle than at the edge cause light to converge on itself; they are known as *convex* lenses. These are some examples:

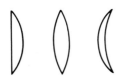

Lenses that are thinner in the middle than at the edge cause light to diverge; they are known as *concave* lenses. These are some examples:

Focus and focal length

a) Convex lenses

Parallel rays of light passed through a convex lens are brought to a focus F.

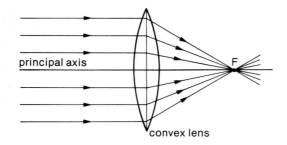

The distance between the centre of the lens and F is the focal length of the lens. A fat lens has a short focal length; a thin lens has a long focal length.

The focus of a convex lens is a real focus because light passes through the focus after it has passed through the lens.

Note that in the above diagram all refraction has been drawn as taking place at a line drawn through the centre of the lens. This is only for convenience: refraction really takes place when light enters and leaves the lens.

b) Concave lenses

Parallel rays of light passed through a concave lens seem to come from a focus F.

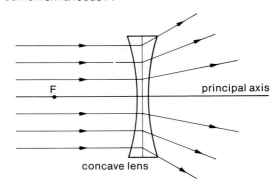

Again the distance between the centre of the lens and the focus is the focal length of the lens. The focus of a concave lens is virtual as light only appears to be travelling from that point after it has passed through the lens.

Image formed by a concave lens

A concave lens only ever produces a virtual, upright image that is smaller than the object. The image is seen by looking through the lens at the object. The nature and position of the image formed by a lens can be described using ray diagrams.

Remembering that a ray shows the direction of the path taken by light, we need only two rays from the object to pinpoint the image:

a) A ray parallel to the principal axis that appears to have come from the focus after refraction through the lens and
b) a ray through the centre of the lens (which does not change direction).

If the diagram is accurately drawn to scale, the size and position of the image is found.

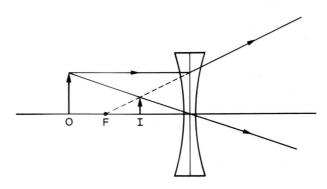

The object is at O, the focus at F and the image at I.

Image formed by a convex lens

The nature and position of the image depends on the position of the object in relation to the focus of the lens.

A ray diagram is drawn in the same way as for a concave lens above, except that the ray parallel to the principal axis passes through the focus after refraction.

a) Object greater than twice the focal length away from lens

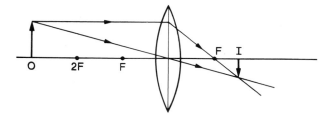

The image is upside down and smaller than the object. The image is created by the intersection of rays and is therefore real; it can be seen by projecting it onto a screen.

b) Object exactly twice the focal length away from the lens

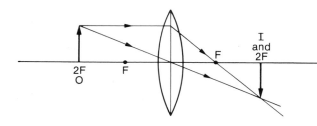

The image is again real and upside down; it is the same size as the object and at twice the focal length on the other side of the lens.

c) Object between F and 2F

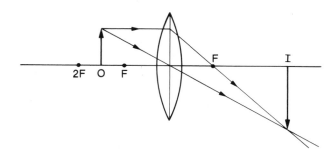

The image is real, upside down and much larger than the object.

d) Object closer to the lens than F

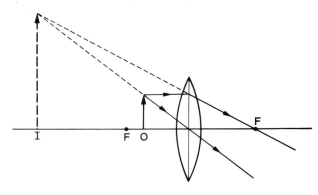

The final image can only be seen by looking at it through the lens; it is a virtual image. It is also larger than the object and upright. The lens is being used here as a magnifying glass.

Finding the focal length of a convex lens

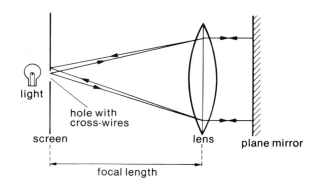

a) Set up the equipment as shown above. The mirror is directly behind the lens.
b) Move the lens and mirror until a clear image of the cross-wires is seen on the screen next to the cross-wires.
c) Measure the distance between the centre of the lens and the screen. This is the focal length of the lens.

When this is done, light leaving the object is refracted by the lens, hits the mirror at right angles and is reflected and refracted back along the same path through the lens to the screen. The parallel rays of light entering the lens after reflection are brought together to form the image at the focus.

Magnification

The magnification M produced by a lens is given by this formula:

$$M = \frac{\text{length of image}}{\text{length of object}}$$

Example

Calculate the magnification if a lens gives an image 3 cm high of an object 2 cm high.

$$M = \frac{\text{length of image}}{\text{length of object}} = \frac{3}{2} = 1.5$$

Magnification is × 1.5.

4.5 The eye

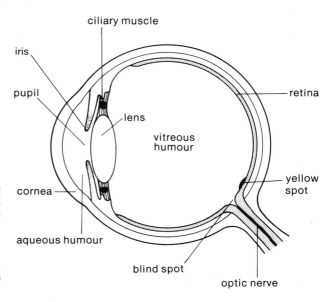

Exercises

1 Draw a diagram showing the action of a) a convex lens and b) a concave lens on a narrow parallel-sided beam of light travelling parallel to the principal axis of each lens. Label each diagram to show:
  i) the principal axis
  ii) the focus
  iii) the focal length
Use the diagrams to explain why the concave lens is called a diverging lens and the convex lens is called a converging lens.

2 Use graph paper to show how a convex lens of focal length 2 cm forms an image of an object 1 cm high placed 5 cm from it. What is the height of the image? How far from the lens is it formed?

3

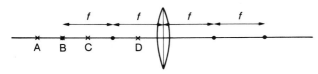

The diagram shows a convex lens of focal length f. An object is placed in turn at A, B, C and D. Which of the descriptions below correspond with the images formed at each object position?
  a) upright, magnified
  b) inverted, magnified
  c) upright, same size as the object
  d) inverted, same size as the object
  e) inverted, smaller than the object
  f) upright, smaller than the object

4 Use a scale drawing to find the position and size of an image of an object 3 cm high placed 15 cm in front of a concave lens of focal length 10 cm.

5 Describe how you would find the focal length of a convex lens.

6 An object 10 cm high is viewed through a convex lens. The image is 15 cm high. What is the magnification achieved by the lens?

The eye detects light and can be thought of as an optical instrument. Light entering the eye is focused by the cornea and the lens; an image of the object that the eye is looking at is cast on the light sensitive retina.

a) Light passes initially through a transparent tissue called the **cornea** and then through a liquid called the aqueous humour. The light is refracted by the cornea, which is part of the focusing mechanism.

b) The **iris** controls the amount of light going further into the eye. The hole at the centre of the iris is the **pupil**; it is opened or closed to let in more or less light depending on the light available.

c) The **lens** refracts the light and focuses it onto the retina. The lens is pliable and its shape (and consequently its focal length) is altered by the action of the ciliary muscles. When looking at a distant object, the focal length is increased and the lens is relatively thin. When looking at a close object, the focal length is decreased and the lens becomes fatter.

The ability of the eye to focus on distant or on close objects is called **accommodation**. The normal eye can accommodate objects that are a very long way away but it cannot accommodate objects nearer to it than a point called the near point. This is about 25 cm for the average person.

d) After the lens, light passes through a fluid called the **vitreous humour** before falling on the retina.

e) The image on the **retina** is upside down. Signals are transmitted to the brain along the **optic nerve** and the brain ensures that the image is seen the right way up.

f) The *yellow spot* is the most sensitive part of the retina and the eye is moved in its socket so as to bring the most important part of the view to focus there.
g) The *blind spot* is the start of the optic nerve and it is not sensitive to light. Light focused on the blind spot is not seen.

Defects of vision

a) *Astigmatism* is caused by an imperfect cornea which does not allow correct focusing and gives rise to a distorted image.
b) A *long-sighted* eye can see distant objects clearly but cannot focus images of close objects. Some people are born long-sighted or it may become a problem in middle-age due to loss of accommodation.

Light from the near point and slightly further away should be brought to a focus on the retina but with a long-sighted eye the focus would be behind the retina.

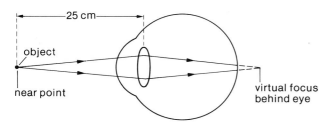

The eyeball is effectively too short.
A *converging* (convex) lens is used in the glasses that correct long sight. This lens, in combination with the eye lens, focuses a clear image onto the retina.

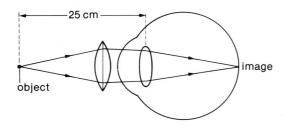

c) A *short-sighted* eye cannot focus images of distant objects clearly.

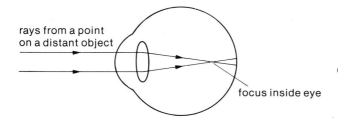

Rays from the distant object are brought to a focus inside the eye and the image on the retina is blurred. The eyeball is effectively too long.
A *diverging* (concave) lens is used in the glasses that correct short sight. A lens of this type, in combination with the eye lens, focuses a clear image onto the retina.

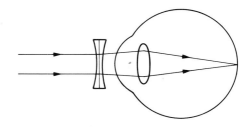

Comparison of the eye and the camera

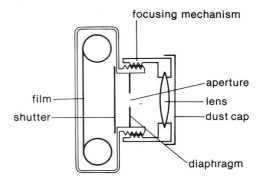

Similarities:
a) Both eye and camera have a convex lens.
b) Light passes through the pupil of the eye and the aperture of the camera. The size of the pupil is controlled by the iris; the size of the aperture is altered using the diaphragm.
c) A real, inverted image that is smaller than the object is focused on light-sensitive film in the camera or on the light-sensitive retina in the eye.

Differences:
a) The lens in the eye changes shape to bring the image to a sharp focus; the lens in the camera is moved either backwards or forwards to effect a focus.
b) The camera has a shutter which opens and closes very quickly and only lets a small amount of light onto the film. This is because the film is changed permanently when light lands on it. The retina can receive and respond to light virtually continuously. There is no equivalent of a shutter in the eye.
c) The dust cap prevents dust in the air from landing on the lens of the camera; the lens has to be cleaned by hand. The eye has an automatic cleaning system with its blinking eyelid and tears.

Exercises

1 Use a labelled diagram of the eye to explain where these structures are and what they do:
   a) the iris
   b) the retina
   c) the optic nerve

2 What is accommodation? A man is looking at a ship on the horizon. Which is the most likely shape of his eye lens from those shown in the diagram below?

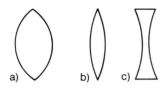

3 What is short sight? Use a diagram to help you to explain how it can be corrected.

4 A woman needs glasses made up from convex lenses to enable her to see clearly. What is the most likely defect in her vision? Why do convex lenses help her?

5 Remember the last time you visited the cinema? Why do we need an usherette to show us to our seats with a torch? Why don't they see us out of the cinema? Why does it always seem to be a very sunny day when you leave the cinema during daylight?

6 List three similarities and three differences between an eye and a camera.

4.6 Optical instruments

Visual angle

The apparent size of an object depends on the size of the image it forms on the retina. The size of the image depends on the size of the *visual angle*:

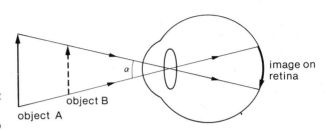

Objects A and B have the same visual angle α but object B is smaller than object A. The images of A and B on the retina are the same size.

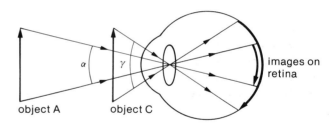

Object C is the same size as object A but because it is nearer the eye it has a larger visual angle γ and a larger image on the retina than A.

Most optical instruments attempt to increase the size of the visual angle so that an object appears to be closer to the eye than it really is.

The magnifying glass

This is the simplest of all optical instruments. The ray diagram was given in section 4.4d (page 40). The object is placed nearer the lens than the focal point and an upright, magnified and virtual image is viewed through the lens. The person using the magnifying glass usually moves it backwards and forwards until the image is at the near point of vision (about 25 cm) from the eye. The magnifying glass increases the visual angle of the object. In the following diagram the visual angle α of the object is shown

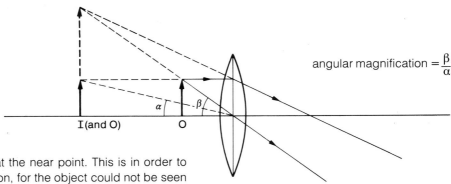

angular magnification = $\frac{\beta}{\alpha}$

with the object also at the near point. This is in order to make a fair comparison, for the object could not be seen clearly without the lens unless it was at least at the near point.

The visual angle of the image is β; β is greater than α and the image is consequently larger than the object.

The astronomical telescope

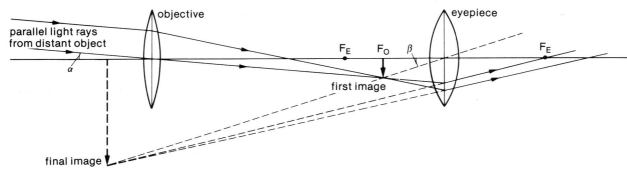

This consists of two convex lenses in a tube. The objective lens is nearer the object and has a larger focal length. The lens nearer the eye is called the eyepiece and it has a short focal length.

The objective forms an upside down and real image of a distant object. The eyepiece acts like a magnifying glass taking the image formed by the objective as its object.

The first image is formed at the focus of the objective F_O. It is closer to the eyepiece than the focus of the eyepiece F_E. The final image is at the near point. It is virtual, upside down and magnified.

The visual angle of the image at the eyepiece is β and the visual angle of the object at the objective is α.

The microscope

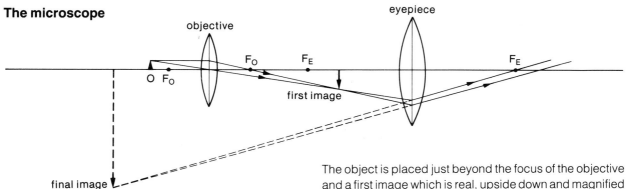

This instrument also has two convex lenses but they both have a short focal length.

The object is placed just beyond the focus of the objective and a first image which is real, upside down and magnified is produced. The eyepiece is then used as a magnifying glass to look at this first image. The final image is even more magnified; it is virtual and upside down.

The reflecting telescope

This has a concave mirror as the objective.

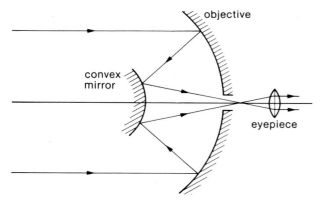

Parallel rays from the distant object are reflected from the objective onto a small convex mirror, then through a hole in the objective and finally into the eyepiece.

Most large telescopes used for serious astronomy are reflecting telescopes because
a) they can be made more easily as there is only one large curved surface to be prepared;
b) they can be supported from behind and they weigh less;
c) there are no problems from unwanted refractions, e.g. the separation of white light into colours (see page 46);
d) it is difficult to make large pieces of glass that are optically perfect throughout.

Slide projector

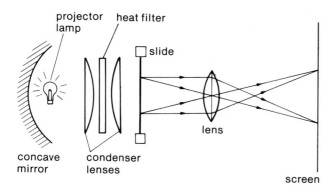

The concave mirror collects light travelling away from the light source and sends it through the condenser which distributes light evenly onto the object slide. Rays of light from two places in the slide have been drawn to show how the image is inverted on the screen.

When the instrument is set up, the position of the lens is adjusted so that the image on the screen is clear. As the instrument is moved further away from the screen the image becomes bigger.

Exercises

1 Explain what is meant by the term 'visual angle'.

2 Draw a diagram to show how a magnifying glass forms an image of an object. Why is it important that the object is close to the glass when it is in use? What sort of an image is formed by a magnifying glass?

3 Draw a diagram to show the arrangement of two lenses being used to make an astronomical telescope. Label the objective and the eyepiece.

Explain why using a telescope helps you to see a distant object more clearly.

4 What kinds of lenses are used in a) a microscope b) an astronomical telescope?

5 Draw a diagram to show how a microscope forms an image of an object.

Why are microscope lenses arranged to give a final image 25 cm from the eye?

6 What kind of telescope does a professional astronomer use and why?

7 What does the condenser do in a slide projector? Why are some slide projectors fitted with fans? What do you have to remember when putting a slide into the projector?

4.7 Colour

The spectrum

Newton showed that white light could be split up into its constituent colours using a triangular glass prism.

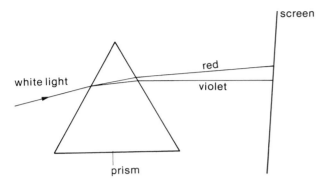

The band of colours produced by the prism is known as the *spectrum* of white light. The colours are red, orange, yellow, green, blue, indigo, violet (in this order). The separation of the colours by the prism is known as *dispersion*.

Why does dispersion occur?

Light is a form of energy and travels in waves. Each colour is associated with light of a particular wavelength. Red light has a longer wavelength than blue light. The amount by which light is refracted on passing into the prism is determined by its wavelength.

red light – longer wavelength – smaller refraction

violet light – shorter wavelength – greater refraction

Therefore, the refractive index of the glass of the prism must be different for each colour.

Recombining the colours of the spectrum

Newton showed that the colours of the spectrum can be recombined by passing them through a second prism. The colours recombine to form white light again.

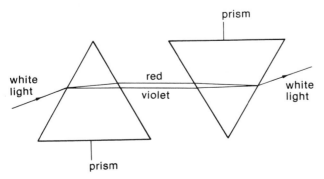

Colours of objects in white light

a) If all colours are reflected the object appears white.
b) If only some colours are reflected, the object appears coloured. Some of the incident light is absorbed and so the object takes in energy and becomes slightly warmer. The colour seen by the eye is the colour of the reflected light.

white objects – reflect all colours

black objects – absorb all colours

red objects – reflect red only and absorb the other colours

Colour filters allow only light of a certain colour to pass through, e.g. a red filter transmits red light and absorbs the other colours.

Appearance of objects in coloured light

To predict the appearance of a coloured object in a coloured light you must consider which colours the object will reflect and which colours it will absorb.

colour of object in white light	colour reflected	colour of light	result of illuminating object	apparent colour of object
red	red	red	red light reflected	red
green	green	red	red absorbed	black
red	red	blue	blue absorbed	black
red	red	green	green absorbed	black

Mixing coloured lights

Red, green and blue are *primary colours* and cannot be formed by mixing lights of any other colour.

Secondary colours can be made by mixing lights of other colours.
 yellow = red + green
 cyan = green + blue
 magenta = red + blue
 red + green + blue = white

Complementary colours give white when mixed together.
 yellow + blue = white
 cyan + red = white
 magenta + green = white

Pigments

An object becomes coloured when a pigment is applied to it. Sometimes artists need to mix pigments to obtain the required shade. A pigment achieves its colour by absorbing (or subtracting) a certain part of the visible spectrum.

A red dress appears red because the dye absorbs all other colours and reflects only red. Yellow paint absorbs blue light and reflects red, yellow and green light. Blue paint absorbs red and yellow but reflects blue and green light. Mixing yellow and blue paint produces green paint because together the yellow and blue absorb red, yellow and blue. This is not the same as mixing yellow and blue light.

Exercises

1 How can you show that white light is a mixture of lights of different colours?

2 The diagram shows a beam of white light striking the side of a prism. Copy the diagram and show how the prism will split up the light into its constituent colours.

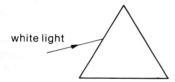

Draw what you would see on the screen AB.

3 What do you understand by the terms   a) primary colour, b) complementary colour?

4 Why does a red object appear red in white light? What happens to its appearance if green light is shone on the object?

5 What is a pigment? Using an example, explain how a pigment achieves its colour.

6 What colours are seen when the following lights are mixed?
   a) yellow and blue
   b) green and blue
   c) red and green and blue
   d) cyan and red
   e) magenta and green
   f) red and green

5 Heat

5.1 Molecules and the kinetic theory

All substances are made up of **atoms**. Atoms join together in groups called **molecules**. An **element** contains atoms of only one kind whereas a **compound** contains at least two different atoms, e.g. iron is an element, sulphur is an element, iron sulphide is a compound of iron and sulphur. One molecule of iron sulphide consists of one atom of iron and one atom of sulphur bonded together.

Atoms and molecules are very small: there are, for example, about 3×10^{23} atoms in 1 g of hydrogen.

The size of a relatively large oil molecule can be estimated with the following experiment.

Measuring the length of an oil molecule

a) Clean a large flat tray and fill it with water.
b) Sprinkle the surface of the water with a fine powder such as lycopodium or talcum.
c) Dip a fine wire into some oil to obtain an oil drop.

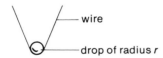

d) Use a magnifying glass and ruler to find the radius r of the oil drop. Make a note of r.
e) Gently place the drop on the surface of the water in the tray. It will spread out to a large circle:

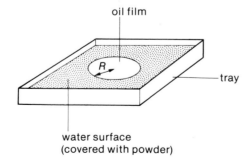

f) Measure and record the radius R of the circle. The volume of the oil drop in $4/3\,\pi r^3$ and the volume of the oil on the water is $\pi R^2 h$, where h is the depth of the oil film.

Oil spreads out on water to form a layer only one molecule thick. This is because one end of the oil molecule is particularly attracted to water. The depth h of the oil layer spread on the water is therefore the length of an oil molecule. When the drop spreads on the water surface, its total volume does not change, so

$$\tfrac{4}{3}\pi r^3 = \pi R^2 h$$

$$h = \frac{4r^3}{3R^2}$$

This experiment gives a value of about 10^{-9} m for the length of an oil molecule.

Forces between two molecules

When two molecules are far apart, they exert no influence on each other. If they are moved closer to each other, two kinds of force begin acting, a repulsive force and an attractive force. When they are a moderate distance apart, the attractive force is bigger than the repulsive one. When they are very close together, the repulsive force is bigger than the attractive one. So there is a certain separation of the molecules for which the repulsive and attractive forces between them are equal. This is known as the equilibrium position. The two molecules settle in the equilibrium position in the absence of other forces, and remain there.

Kinetic theory

This is the theory that all atoms and molecules of matter, in whatever state, are moving continuously. The energy of the movement of the molecules in any object is the **heat energy** of that object. As the object's temperature increases, the movement of its molecules becomes more violent. Evidence of molecular movement comes from experiments on **diffusion** and on **Brownian motion**.

Diffusion

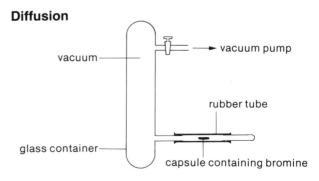

The capsule containing the liquid bromine is crushed by gripping the rubber tube in a pair of pliers. The container is filled almost instantaneously with brown bromine gas. The molecules of bromine move very rapidly. If air is allowed to remain in the container, the bromine gas still diffuses into the container but much more slowly. The molecules of bromine move rapidly but undergo many collisions with the air molecules.

Brownian motion can be demonstrated with the smoke cell.

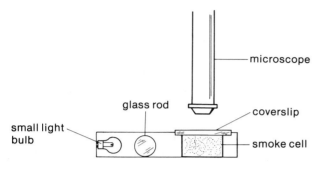

The cell is filled with smoke from a burning straw, closed with a cover slip and then placed under the microscope. The cell is illuminated from the side by a small bulb; the light passes through a glass rod, acting as a lens on its way to the smoke. The smoke particles scatter some of the light up through the microscope. When viewed through the microscope, the particles of smoke seem to be moving all the time, first in one direction and then in another. This can be explained if it is imagined that these smoke particles are being hit all the time by air molecules coming at them in all directions.

This type of movement was first noticed by Robert Brown in the last century and it is named after him. He was looking at pollen grains on water. The grains were moving on the water surface because they were being bombarded by water molecules. This effect is now called Brownian motion.

Solids, liquids and gases

There are three states of matter: solid, liquid and gas.

a) Solids

In a solid, the molecules are arranged in a definite pattern; they are held fairly rigidly together and this gives the solid a definite shape. The molecules are moving all the time, vibrating about their equilibrium positions. The hotter the solid the more energetic are the vibrations.

b) Liquids

In a liquid, the molecules are not arranged in any particular pattern. The molecules are much freer to move with respect to each other and so liquids do not have any fixed shape.

c) Gases

In a gas, the molecules can move completely free of each other. They are widely separated and move around quickly. Because gas molecules exert very little influence on each other they can easily move apart and so a gas is free to expand to fill any empty space into which it is introduced.

The pressure of a gas is the force exerted on each square metre of the sides of the container by the gas as a result of the collisions of the moving gas molecules with the sides.

Evaporation

The molecules of a liquid are constantly moving and exchanging energy. Some molecules have more energy than others. If they happen to be at the surface, some molecules may have enough energy to leave the liquid completely. This is called *evaporation*. For instance pools of rain water lying on the pavement eventually evaporate and the water disappears.

Evaporation is more noticeable in some liquids than it is in others. The rate of evaporation of a liquid can be increased by:
a) increasing the temperature of the liquid. If the energy of the liquid is increased its molecules move more quickly and more molecules have enough energy to escape.
b) increasing the surface area of the liquid. This increases the chance of a molecule being at the surface.
c) Placing the liquid in a draught. The air carries away the molecules that have left the liquid so that they cannot return.

During evaporation, molecules with a lot of energy leave the liquid and therefore the average energy of those left is lower. Thus evaporation causes cooling of the liquid.

Surface tension

A needle can be made to float on the surface of water – look closely at this diagram and you will see that the needle seems to be supported by an elastic surface skin.

This effect is called *surface tension*. The molecules in the surface are not surrounded on all sides by neighbours and have unbalanced forces acting on them. The surface molecules are pulled in towards the main bulk of the liquid and this creates the surface tension. The surface tension of the liquid can be changed by adding another substance to it, e.g. a detergent lowers the surface tension of water. This enables the water to penetrate into the fibres of dirty clothes to clean them more effectively. It also means that bubbles last longer before bursting.

Exercises

1. An oil drop of radius 1 mm forms a circle of radius 0.9 m on the surface of water. If the oil spreads to make a layer one molecule thick, estimate the length of an oil molecule.

2. What forces act between molecules? What is meant by equilibrium position?

3. What is the kinetic theory? How is it used to help explain the arrangement of molecules in solids, liquids and gases?

4. What evidence is there to suggest that the kinetic theory might be right?

5. If a few blue crystals of copper sulphate are left in the bottom of a beaker of water they gradually dissolve and colour all of the liquid blue. What is this an example of? Explain what is happening using the kinetic theory.

6. What happens when a liquid evaporates? How can the rate of evaporation of a liquid be increased? What are the ideal weather conditions for drying wet clothes?

7. Why can a needle be made to float on water? Why does it sink when a few drops of alcohol are added to the water?

8. How can gas pressure be explained with the kinetic theory? What would you expect to happen to the gas pressure if the temperature was increased?

5.2 Thermometry

Heat and temperature

The *temperature* of an object is a measure of the kinetic energy of its molecules. If the object's temperature is increased, then its molecules move more quickly. If two objects having different temperatures are brought together, energy from the hotter object then flows to the cooler object. We call this energy *heat*.

When the two objects have been in contact for a while they end up at the same temperature; there is no net heat flow from one to another; they are in thermal equilibrium.

Temperature scales

Various scales of temperature have been devised to give numerical values to temperature. The principles of temperature scales are the same for every scale:

a) A substance with a physical property which varies uniformly over the temperature range desired is chosen, e.g. the expansion of a liquid. The liquid is known as the *thermometric substance*.

b) The values of the property at two easily reproducible temperatures, called fixed points, are measured. These values represent the limits of the scale.

c) The interval between the two fixed points is divided up into a number of equal parts (sometimes called degrees).

The Celsius (centigrade) scale of temperature

The *lower fixed point* is the temperature of pure melting ice. The purity is essential because impurities lower the melting point; the ice must be melting, otherwise it would be at a temperature lower than the melting point.

This fixed point is known as 0°C.

The *upper fixed point* is the temperature of steam above water boiling under standard atmospheric pressure (1.01×10^5 N/m²). The pressure must be at this value because the temperature of steam above boiling water varies with pressure. It does not vary with impurity in the water.

This fixed point is known as 100°C.

The temperature difference between these fixed points is divided up into 100 equal degrees Celsius.

Calibrating a thermometer

The two fixed points can be checked on a calibrated thermometer or put on a new, uncalibrated thermometer in the following ways.

a) Upper fixed point

Water is boiled in a hypsometer; the thermometer bulb is surrounded by pure water vapour at atmospheric pressure. The manometer shows that the pressures inside and outside the hypsometer are the same. The steady level reached by the mercury in the thermometer is marked as the upper fixed point (100°C).

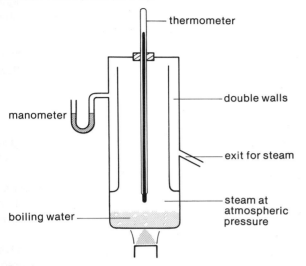

b) Lower fixed point

The thermometer is surrounded by pure melting ice. Again the steady level reached by the mercury is marked on the stem as the lower fixed point (0°C).

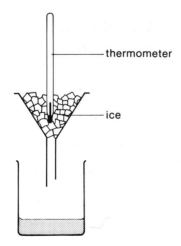

Liquid-in-glass thermometers

Many different physical properties may be used in thermometers: expansion of liquids, solids and gases; electrical resistance; thermal electron moving force and so on. The most common thermometer uses the expansion of a liquid in a glass tube as the thermometric property. These thermometers are made from a length of capillary tubing sealed at one end, connected to a thin-walled glass bulb. This contains liquid, usually mercury or alcohol. When the temperature of the bulb goes up the mercury expands up the tube. The length of the mercury column depends on the temperature of the bulb. The important features of the design of this very successful thermometer are labelled on the diagram below.

A liquid-in-glass thermometer

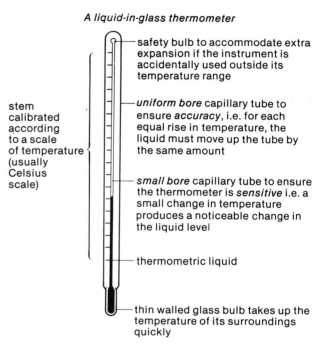

Mercury or alcohol?

These two are the most common liquids in liquid-in-glass thermometers and each one has certain advantages over the other.

Mercury is useful over a larger range: mercury melts at −39°C and boils at 357°C; alcohol melts at −117°C and boils at 78°C.

Alcohol is used for low temperature work; mercury for higher temperature measurement.

Mercury is easy to see and it does not wet the glass; alcohol has to be dyed, as it is normally colourless, and it does wet the glass. Some droplets of alcohol can be left behind sticking to the tube as the temperature falls.

Both mercury and alcohol expand uniformly but alcohol expands more per degree than mercury (about six times as much). Alcohol is thus more sensitive to temperature changes. However, mercury, being a metal, conducts heat much more quickly than alcohol so mercury responds more quickly to temperature changes.

The *clinical thermometer* is a special liquid-in-glass thermometer.

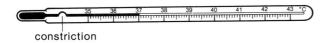
constriction

This thermometer is specially designed to measure human body temperature. It measures over a very limited range of 35–43°C. 'Normal' body temperature is 37°C. The thermometer contains mercury and is usually placed either under the tongue or in the armpit and is left in position for a minute or two before being read.

Before use, the thermometer must be shaken so that the mercury level is below the constriction. The constriction is a narrowing in the capillary tube. When the thermometer is in the mouth the force of expansion pushes the mercury up the tube. On removing the thermometer from the mouth, the constriction prevents the mercury in the stem of the thermometer from returning to the bulb. This enables an accurate reading of the body temperature to be taken after the thermometer has been removed from the mouth.

Exercises

1 Explain the difference between heat and temperature. Which has more heat energy, the sea or a red-hot iron bar? Which way does the energy flow when the bar is plunged into the sea?

2 What do you need in order to set up a temperature scale? Why should the fixed points of the scale be reproducible all over the World?

3 Describe how you would calibrate an unmarked thermometer. How would you measure an unknown temperature that was somewhere between 0°C and 100°C?

4 What are the advantages in using alcohol in a liquid-in-glass thermometer compared to mercury? Which thermometer would you take on an Antarctic expedition, alcohol or mercury?

5 What features of a mercury thermometer make it especially sensitive to temperature changes?

6 In a clinical thermometer, explain why
   a) a narrow range of about 10°C is sufficient.
   b) there is a constriction.
Why should a clinical thermometer never be sterilised in boiling water?

7 Why are these features essential in the design of a mercury-filled thermometer?
   a) the capillary tubing has a narrow bore
   b) the capillary tubing has a uniform bore
   c) the bulb containing the mercury has thin walls

5.3 Expansion of solids and liquids

Expansion of solids

When a bar of metal is heated, it expands and becomes longer. From experiments it is found that the amount of expansion depends upon the temperature rise. If the temperature decreases, then the bar will contract (shrink).

Very great forces are set up when materials expand and contract and considerable damage may be done if allowance is not made for these effects in many circumstances.

a) Large areas of concrete can expand and contract even with typically British weather. The concrete is laid in sections about 10 m wide with a thin layer of pitch in between sections. When the temperature rises and the concrete expands the pitch is forced out, to return when the temperature drops. Without the pitch, the concrete may crack during such temperature changes.

b) Railway tracks are always laid so that they do not buckle when they expand. Joints between sections of track are overlapped.

c) Bridges expand considerably on hot summer days. This is allowed for with small gaps, one at each end, covered by metal plates. The bridge is supported on rollers which can move.

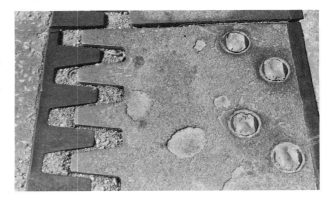

The *expansivity* of a solid bar is the fraction of the original length by which the bar expands on heating through one kelvin (1 K): see page 55. Expansivity can be calculated using this equation

$$\text{expansivity} = \frac{\text{increase in length of bar}}{\text{original length} \times \text{temperature rise}}$$

The expansivity of a solid is a constant; it is the same regardless of size of bar or temperature rise.

Here are some typical values of expansivity:

brass	0.000019 per kelvin
aluminium	0.000026 per kelvin
glass	0.000008 per kelvin
concrete	0.000011 per kelvin

If, for example, an aluminium bar one metre long is heated up by one kelvin, then its length will increase by 0.000026 m. If the temperature rise is 10 K, then the expansion will be 0.00026 m and so on.

Measurement of expansivity

The expansivity of a metal bar is determined with equipment similar to that shown below.

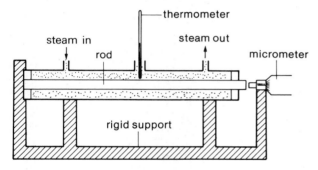

The length of the rod is measured before it is placed in the apparatus and, when it is in place, its temperature is taken. The micrometer is tightened up and a reading taken from it. Then it is slackened off to allow for the expansion. The temperature of the rod is then raised by passing steam around it. After a while the micrometer is tightened up and a new reading taken. The difference in the two micrometer readings is the amount of expansion of the rod. This final step is usually repeated after a few more minutes to make sure that the rod has entirely reached its maximum temperature and that the expanding has finished. A final temperature reading is taken and the expansivity calculated as above.

Uses of expansion in solids

a) **Fitting of steel tyres**

The tyre is heated strongly and then fitted on to the wheel. When the tyre cools it contracts onto the wheel to fit tightly. Axles are often fitted to wheels in a similar fashion. The axle is dipped in liquid nitrogen; it contracts and is then pushed into the hub. When it warms up to room temperature it expands to fit tightly.

b) **The bimetallic strip**

A bimetallic strip consists of two metals (such as brass and iron) of differing expansivities joined together. When the strip is heated, it bends because the two metals expand by differing amounts.

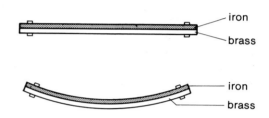

The bimetallic strip can be used in a thermostat. As the temperature increases the strip bends and eventually breaks electrical contact in the heater circuit. When the temperature decreases, the bimetallic strip returns to its original position and shape and the contact is restored.

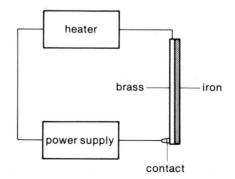

Expansion of liquids

Virtually all liquids expand when they are heated. Some expand more than others and this variation can be shown with this equipment:

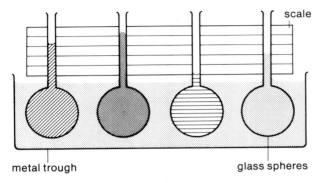

The spheres contain a variety of liquids and they are all filled to the same level. Boiling water is poured into the metal trough; the liquids warm up and expand. The amount of expansion of each liquid is clearly visible against the background of the scale.

Uses of expansion in liquids are mainly in thermometry, discussed on page 51.

The expansion of water

As water is heated from 0°C to 4°C it actually contracts. It expands as it is heated above 4°C. This means that any mass of water takes up the smallest space at 4°C; water is at its most dense at 4°C.

As the water in a pond cools during winter the most dense water, at 4°C, sinks to the bottom of the pond and warmer water from the bottom takes its place. On further cooling the most dense water, at 4°C, stays at the bottom because water above it, cooling from 4°C down to 0°C, is less dense. Eventually ice forms on the top of the pond. Further heat loss continues much more slowly, so in deep ponds there is almost always unfrozen water at the bottom.

Exercises

1 Give two uses and two disadvantages of expansion in metals.

2 a) By how much would a three metre rod of aluminium expand when heated by one kelvin?
b) By how much would a two metre bar of brass contract when cooled by five kelvin?
(Use values for expansivity on p. 53)

3 Outline an experiment to find the expansivity of iron. Calculate the expansivity of iron from the data shown below.
Original length of bar = 0.6 m
Final length of bar = 0.600576 m
Original temperature of bar = 20°C
Final temperature of bar = 100°C

4 Explain why a bimetallic strip bends when it is heated. Use a bimetallic strip to design a fire alarm.

5 Here is a photograph of a long pipe carrying hot oil in an oil refinery.

You will see that the pipe has been looped up in the air and down again: why is this so?

6 The metal lids on glass jars of food sometimes stick, particularly when new. You can free the lid by dipping it in hot water for a few moments. How does this work?

5.4 The gas laws

The volume of a gas may be altered by changing its pressure, or its temperature, or both.

Charles' law Any fixed mass of gas increases in volume by 1/273 of its volume at 0°C for every increase in temperature of 1°C, providing the pressure remains constant.

This relationship can be shown to be true with the equipment illustrated.

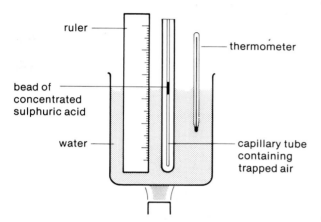

The gas is trapped in the capillary tube by the bead of concentrated sulphuric acid. The acid absorbs any moisture in the gas. The volume of the gas at any temperature is found by multiplying the length of the column of trapped gas by its cross-sectional area. The temperature of the gas can be altered by heating the water bath. The gas remains at atmospheric pressure because the capillary tube is open to the air.

A graph of the results clearly shows that the volume of the gas is proportional to its temperature.

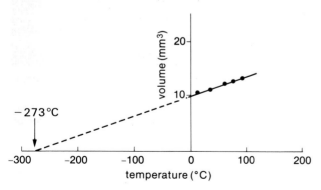

The thermodynamic scale of temperature

If the graph above is extrapolated to lower temperatures, it crosses the temperature axis at −273°C. At this temperature, the volume of any gas would theoretically be zero. It would seem sensible to take this temperature as the true zero of temperature and to start a temperature scale there. This has been done and the scale is known as the thermodynamic scale of temperature. The word 'thermodynamic' refers to the movement of heat energy and is used in this instance because at absolute zero the heat energy content of anything is also zero.

The unit of temperature on this scale is the kelvin (symbol, K) and it has the same size, exactly, as one degree Celsius. So 0°C is equivalent to 273 K, −273°C is 0 K and 100°C is 373 K, etc.

Converting from the Celsius to the thermodynamic scale

a) To change from °C to K you must add 273.

0°C = 273 K
20°C = 20 + 273 = 290 K
−250°C = −250 + 273 = 23 K

b) To change from K to °C you must subtract 273.

0 K = −273°C
15 K = 15 − 273 = −258°C
420 K = 420 − 273 = 147°C

Pressure law The pressure of any fixed mass of gas increases by 1/273 of its pressure at 0°C for every increase in temperature of 1°C, providing the volume is kept constant.

This relationship can be shown to be true with the equipment illustrated.

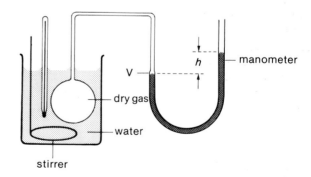

If the right-hand mercury level is higher than the left, the pressure of the gas is greater than atmospheric.

If the right-hand mercury level is lower than the left, the pressure of the gas is less than atmospheric.

The pressure of the gas is the difference in the mercury levels h plus atmospheric pressure. If the mercury levels are the same, the pressure of the gas is atmospheric.

The volume of the gas is kept constant by raising or lowering the manometer to keep the mercury at the mark V while the experiment proceeds. The temperature is raised by warming the water and is read from the thermometer.

The results are similar to the Charles' law experiment.

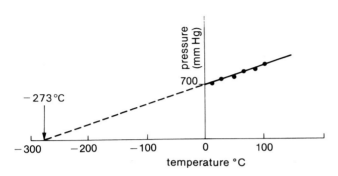

Boyle's law The pressure of any fixed mass of gas is inversely proportional to its volume, providing the temperature remains constant.

This relationship can be shown to be true with the equipment illustrated.

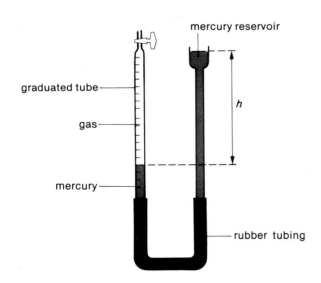

The volume of the trapped gas is read from the graduations on the left-hand tube. The pressure can be varied by raising or lowering the reservoir; the actual pressure of the gas is the difference in the mercury level h plus atmospheric pressure.

A graph of pressure against volume at constant temperature looks like this:

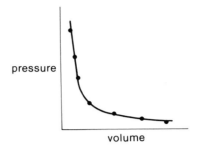

A graph of pressure against 1/volume at constant temperature is shown below.

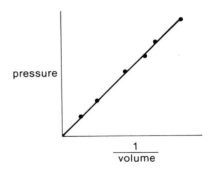

The information in the graph is brought together as Boyle's law.

Worked examples of problems

a) Charles' law

A certain mass of gas takes up 1.5 m³ at 27°C. It is heated at constant pressure until it takes up 2 m³. What is its new temperature?

This problem is solved by using the fact that the volume V of a gas is proportional to its temperature T in kelvin.

$$V \propto T$$
$$\text{or } \frac{V}{T} = \text{a constant}$$

Taking $V_1 = 1.5$ m³ $T_1 = (27 + 273)$ K
       $V_2 = 2$ m³ $T_2 = $ unknown

$$\frac{V_1}{T_1} = \frac{V_2}{T_2}$$

$$\frac{1.5 \text{ m}^3}{(27 + 273) \text{ K}} = \frac{2 \text{ m}^3}{T_2}$$

$$T_2 = 400 \text{ K } (127°C)$$

b) Pressure law

A certain mass of gas at 110 000 N/m² pressure is sealed into a jar at 20°C. The jar is heated to 50°C. What will be the new pressure of the gas?

The pressure of the gas P is proportional to its temperature T in kelvin.

$$P \propto T$$

$$\text{or } \frac{P}{T} = \text{a constant}$$

Taking $P_1 = 110\,000$ N/m² $T_1 = (20 + 273)$ K
$P_2 = \text{unknown}$ $T_2 = (50 + 273)$ K

$$\frac{P_1}{T_1} = \frac{P_2}{T_2}$$

$$\frac{110\,000 \text{ N/m}^2}{(20 + 273) \text{ K}} = \frac{P_2}{(50 + 273)}$$

$$P_2 = 121\,000 \text{ N/m}^2$$

c) Boyle's Law

A bubble contains 50 cm³ of helium at 120 000 N/m² and it is allowed to expand to a new volume when the pressure is decreased to 100 000 N/m². What is its new volume?

The pressure of the gas P is inversely proportional to the volume V.

$$P \propto \frac{1}{V}$$

$$\text{or } PV = \text{a constant}$$

Taking $P_1 = 120\,000$ N/m² $V_1 = 50$ cm³
$P_2 = 100\,000$ N/m² $V_2 = \text{unknown}$

$$P_1 V_1 = P_2 V_2$$

$$120\,000 \text{ N/m}^2 \times 50 \text{ cm}^3 = 100\,000 \text{ N/m}^2 \times V_2$$

$$V_2 = 60 \text{ m}^3$$

Note: In some calculations it may be necessary to convert pressure readings in mm mercury to N/m² or vice versa. How to do this is shown on page 5.

Calculations can be solved using either mm mercury or N/m², but care must be taken not to mix up these units in any one calculation.

Exercises

1. a) State Charles' law.
   b) Draw a diagram of the apparatus you would use to show that Charles' law is true.
   c) How do the results of the Charles' law experiment lead to a temperature scale based on an absolute zero of temperature of −273°C?

2. Convert a) −196°C, b) 10°C and c) 1068°C into kelvin.
   Convert d) 5 K, e) 600 K and f) 1068 K into °C.

3. State the pressure law.
   A certain mass of gas is heated from 0°C to 273°C. What happens to its pressure?

4. a) State Boyle's law.
   b) Draw a diagram of the apparatus you would use to show that Boyle's law is true.
   c) Explain how:
      the volume of the gas under investigation is measured;
      the pressure of the gas under investigation is measured.
   d) Sketch the graph you would expect to draw using your results.

5. 2 m³ of hydrogen at a pressure of 100 000 N/m² are compressed to a volume of 1 m³. What is the new pressure of the gas if its temperature remains constant?

6. 2 m³ of hydrogen at a pressure of 100 000 N/m² and a temperature of 25°C are heated to 75°C. If the volume is kept at 2 m³, what will be the pressure at 75°C?

7. 4 m³ of helium at 10°C are heated to 60°C and the pressure remains constant. What will be the volume at 60°C?

8. Design a thermometer that uses the expansion of air to measure temperature. Explain why it is unlikely to be very accurate.

5.5 Heat measurement

Specific heat capacity

All forms of energy are measured in *joules* (J). Heat (thermal energy) is no exception.

When joules of heat are given to an object, its temperature goes up. We say that the object has been heated. The size of the temperature rise depends upon the mass of the object.

The number of joules of heat required to raise the temperature of an object by one kelvin is known as the *heat capacity* of the object. An object with a large heat capacity requires a large number of joules to raise its temperature by one kelvin.

The *specific heat capacity* of a substance is the number of joules of heat required to raise the temperature of one kilogram of the substance by one kelvin.

Heat capacity C has units of J/K.

Specific heat capacity c has units of J/(kg K).

Here are some examples of specific heat capacities, c.

Substance	c J/(kg K)	Substance	c J/(kg K)
aluminium	900	lead	130
brass	380	mercury	140
copper	400	brine	3900
glass	470	water	4200
ice	2100	zinc	380
iron	460	methylated spirit	2400

Measuring specific heat capacity for a solid can be done in the following manner.

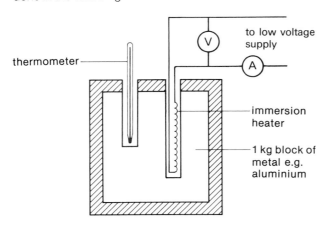

a) Set up the 1 kg block with the immersion heater and thermometer in place and connect the circuit. Surround the block with insulation to cut down on heat loss to the surroundings.
b) Record the temperature, then switch on the heater for a known time t seconds. Note the voltmeter reading V volts, and the ammeter reading, I amps, and adjust the power supply if necessary so that V and I remain constant during t.
c) Switch off after t seconds and note the highest temperature reached by the block. Calculate the temperature rise θ.

The heat, in joules, supplied by the heater to the block in t seconds is

$$(I \text{ amps} \times V \text{ volts} \times t \text{ seconds})$$

This gave a temperature rise of θ K and the specific heat capacity c is for a rise of 1 K, therefore

$$c = \frac{I \times V \times t}{\theta}$$

For a block of mass m kg, the equation would be

$$c = \frac{IVt}{m\theta}$$

Specific heat capacities for liquids are measured in a similar way.

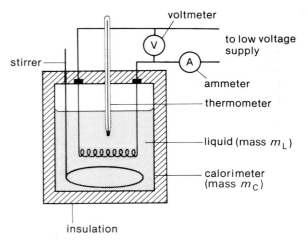

The liquid has mass m_L; the calorimeter has mass m_C. A calorimeter is a metal beaker of known specific heat c_C especially designed for experiments on heat capacities.

The heater warms up the liquid and the calorimeter, and an allowance must be made for the heat taken by the calorimeter.

The energy supplied by the heater in t seconds is

$$(I \times V \times t) \text{ joules.}$$

The energy taken by the calorimeter for a temperature rise of θ K is $(m_C \times c_C \times \theta)$ joules, where m_C is the mass in kg and c_C is the specific heat capacity in J/(kg K) of the calorimeter. The remaining energy taken up by the liquid for an equal temperature rise of θ K is $(m_L \times c_L \times \theta)$ where m_L and c_L are the mass in kg and specific heat capacity in J/kg K of the liquid respectively.

$$\frac{\text{energy from}}{\text{heater}} = \frac{\text{energy into}}{\text{liquid}} + \frac{\text{energy into}}{\text{calorimeter}}$$

$$(I \times V \times t) = (m_L \times c_L \times \theta) + (m_C \times c_C \times \theta)$$

$$c_L = \frac{(I \times V \times t) - (m_C \times c_C \times \theta)}{m_L \theta}$$

Worked examples

a) How much heat energy is needed to raise the temperature of 2 kg of aluminium by 10°C if the specific heat capacity is 900 J/(kg K)?

heat required $= m \times c \times \theta$
$= 2 \text{ kg} \times 900 \text{ J/(kg K)} \times 10 \text{ K}$
$= 18\,000 \text{ J}$

b) How much heat energy is given out when a 3 kg brass block cools by 5°C if the specific heat capacity of brass is 380 J/(kg K)?

heat liberated $= m \times c \times \theta$
$= 3 \text{ kg} \times 380 \text{ J/(kg K)} \times 5 \text{ K}$
$= 5700 \text{ J}$

c) In an experiment, a 1 kg block of copper was heated for 1000 seconds using an immersion heater. The voltmeter and ammeter read 12 V and 1 A, respectively, and the temperature rise recorded was 25°C. Calculate the specific heat capacity for copper.

$m \times c \times \theta = I \times V \times t$

where m is the mass of the calorimeter $= 1$ kg
c is the specific heat capacity $=$ unknown
θ is the temperature rise $= 25°C = 25$ K
I is the current supplied $= 1$ A
V is the potential difference $= 12$ V
t the time in seconds $= 1000$ s

$1 \text{ kg} \times c \text{ J/(kg K)} \times 25 \text{ K} = 1 \text{ A} \times 12 \text{ V} \times 1000 \text{ s}$

$$c \text{ J/(kg K)} = \frac{1 \text{ A} \times 12 \text{ V} \times 1000 \text{ s}}{1 \text{ kg} \times 25 \text{ K}}$$

$c = 480$ J/(kg K)

Exercises

(Use values for specific heat capacities given on page 58 where necessary.)

1 What is meant by the 'heat capacity' of a substance? How does it differ from the 'specific heat capacity'?

2 How much heat is needed to raise the temperature of 1 kg of brass by 10°C?

3 By how much will the temperature of 5 kg of iron rise if it is given 345 kJ of heat?

4 How much heat is needed to raise the temperature of 10 kg of ice from $-12°C$ to $-7°C$?

5 How much heat is lost by 4 kg of water cooling from 90°C to 20°C?

6 In an experiment to measure the specific heat capacity of brass by an electrical method the following readings were obtained:

voltage across the heating coil $= 4.5$ V
current through the heating coil $= 0.75$ A
current flowed for 480 seconds
mass of brass heated $= 0.5$ kg
temperature rise $= 7.5°C$

Calculate a value for the specific heat capacity of brass from these readings.

Compare this value with the value given on page 58.

What changes could be made to the experiment to improve the accuracy of the readings?

7 In an electrical method to measure the specific heat capacity of brine the following measurements were made:

mass of aluminium calorimeter $= 0.1$ kg
mass of brine $= 0.05$ kg
temperature change $= 7.5°C$
voltage across the heating coil $= 5$ V
current through the heating coil $= 1$ A
current flowed for 450 seconds

Calculate a value for the specific heat capacity of brine from these results, given that the specific heat capacity of aluminium is 900 J/(kg K).

5.6 Change of state

Matter can exist in three different states: solid, liquid and gas. Changes of state occur as a result of changes in temperature and/or pressure.

When a solid changes into a liquid, we say that it has *melted* or fused. Melting occurs at a fixed temperature for a pure substance. The melting point of a pure substance is lowered by the addition of dissolved impurities.

When a liquid changes into a solid, it has frozen or *solidified*.

Evaporation and boiling

A liquid evaporates when it changes into a gas. The opposite of evaporation is condensation. *Evaporation* takes place at the surface of the liquid over a range of temperatures. As a liquid is heated, the evaporation rate is increased because more molecules in the liquid have enough energy to escape.

Boiling occurs when molecules are able to escape from all parts of the liquid. Bubbles of gas form anywhere in a boiling liquid and rise to the surface. Boiling happens at one fixed temperature for a given atmospheric pressure. In fact, the boiling point is reached when the pressure created by the evaporating gas is exactly equal to atmospheric pressure.

Pressure and boiling

A liquid boils at a temperature lower than expected if the pressure above it is reduced.

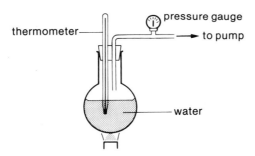

If a vacuum pump is used to draw air out of the flask, the water boils even though the thermometer registers a temperature below 100°C.

When the pressure above a liquid is increased, the liquid boils at a higher temperature. This fact is used in the pressure cooker. Food cooks much more quickly in a pressure cooker because the boiling point of water is raised to about 105°C by allowing the pressure inside the cooker to increase.

Pressure and melting

An increase in pressure can result in a lowering of the melting point for some solids. It happens to those solids whose volume decreases on melting, e.g. water.

The diagram shows a block of ice with a wire carrying weights around it.

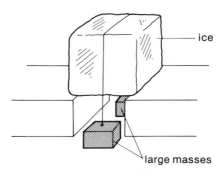

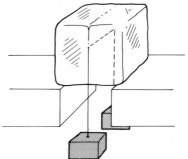

The wire passes slowly through the ice block, leaving it completely solid. The pressure under the wire is sufficient to lower the melting point; the ice melts and the wire sinks through the melted water which then re-freezes.

Dissolved substances raise the boiling point and lower the freezing point of a liquid.

Salt is spread on the roads in winter to lower the freezing point and keep the roads free of ice.

Cooling by evaporation

a) A beaker of ether can be made to evaporate by blowing air through it. If the beaker is standing in a small amount of water, the water will freeze. As the ether evaporates, it takes heat energy from the water.

b) People perspire after exercise; the sweat evaporates and helps to cool the body.

c) Refrigerators rely on the evaporation of volatile liquids to keep food cool.

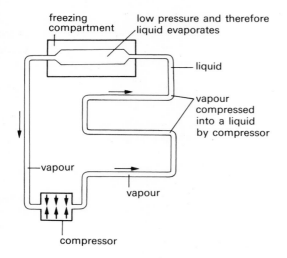

The liquid evaporates within the metal skin of the freezing compartment, taking energy from the inside of the refrigerator to do so. The food inside the refrigerator is cooled.

The vapour is pumped around in the pipes and is condensed by the compressor. This is usually situated at the back of the refrigerator and is outside because it gets warm while running.

Heating and cooling curves

If a sample of ice at a temperature well below 0°C is heated slowly and its temperature is recorded at regular time intervals, a graph similar to this one may be drawn from the results.

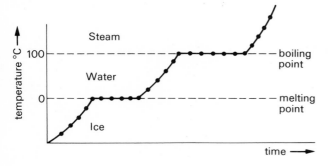

Notice that at the melting point and boiling point, the temperature remains constant despite the fact that the ice or ice-water is being heated. The heat energy that is taken by a substance to enable it to change state, and that is not registered as a temperature change, is called *latent heat*.

The same shape of graph is given in reverse when the steam condenses and is cooled until it solidifies.

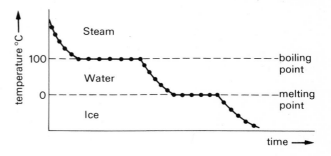

The latent heat is given out as the water changes state and this maintains the temperature at the boiling or melting point until the change is complete.

The specific latent heat of vaporization is the heat needed to change one kilogram of the substance from liquid to vapour at the boiling point. It is measured in J/kg. The specific latent heat of vaporization can be measured in the following way:

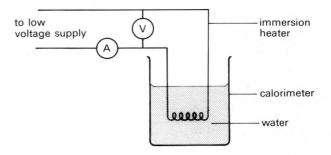

a) Find the mass of the water in the calorimeter and heat it until it boils.
b) As soon as the water boils, start a clock and continue heating for a time t seconds.
c) Record the ammeter and voltmeter readings.
d) Remove the heater after t seconds and find the mass of the water m that has evaporated.

The energy supplied by the heater in joules =
V volts × I amps × t seconds

and this is sufficient to evaporate m kilograms of water.

So the specific latent heat of vaporization of water L is given by

$$L = \frac{V \times I \times t}{m} \text{ J/kg}$$

For water, the specific latent heat of vaporization is 2 260 000 J/kg.

The specific latent heat of fusion is the heat needed to change one kilogram of a substance from solid to liquid at the melting point. It is measured in J/kg.

The specific latent heat of fusion of water can be measured in the following way:

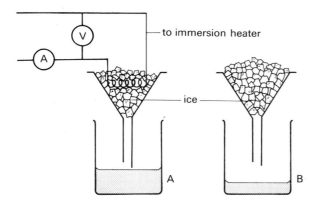

a) Fill two identical filter funnels with the same mass of crushed ice and place each over a beaker.
b) Put an immersion heater into one and pass a current through it for a time t seconds.
c) Record the ammeter and voltmeter readings.
d) After t seconds find the mass of melted water in each beaker. Subtract the mass of water in beaker B from that in beaker A. The water in B is the amount of ice that is melted by the heat of the room during the course of the experiment. The calculation should only include the mass of ice m melted by heat from the immersion heater.

As above, the specific latent heat of fusion L of water is calculated from

$$L = \frac{V \times I \times t}{m}$$

For water, the value of L is 336 000 J/kg.

Exercises

1. Explain what is meant by a) fusion, b) evaporation.
2. What is the difference between evaporation and boiling?
3. Describe an experiment to show that boiling can occur at temperatures below those expected if the pressure above the liquid is reduced.
4. What is the effect of pressure on the melting point of water?
5. Explain what is meant by these terms.
   a) specific latent heat of vaporization
   b) specific latent heat of fusion
6. How much heat is needed to melt 1 kg of ice at 0°C?
7. How much heat is needed to evaporate 1 kg of water at 100°C?
8. 2 kg of water at 100°C is obtained when steam condenses. How much heat is given out in this change?
9. How much heat is needed to turn 2 kg of ice at −2°C to water at 2°C? Values of specific heat capacities needed are given on page 58.
10. How much heat is given out when 0.5 kg of steam at 100°C condenses to water and cools to 80°C?
11. How much heat is needed to turn 0.75 kg of ice at −10°C to steam at 100°C?
12. Sketch a cooling curve for a liquid in the range 20°C to 80°C if the liquid freezes at 30°C.
13. These results were obtained in an experiment using an electrical method to measure the specific latent heat of fusion L of ice. Calculate L and comment on the result.

   voltage across heating coil = 2 V
   current through the coil = 2 A
   mass of ice melted = 0.005 kg
   current flowed for 450 seconds

14. Calculate the specific latent heat of vaporization of water using these results obtained using an electrical method.

   voltage across the heating coil = 12 V
   current through the coil = 3 A
   mass of water evaporated = 0.0095 kg
   current flowed for 600 seconds

5.7 Heat transmission

Heat can travel in three ways: conduction, convection and radiation. Atoms and molecules take part in the processes of conduction and convection, but radiation can occur through a vacuum.

Conduction

The ability of a material to conduct heat is called its conductivity. In general, metals conduct well and non-metals are not usually good conductors. Low conductivity materials are used as insulators for lagging. The thermal conductivity of different solids can be compared using this apparatus.

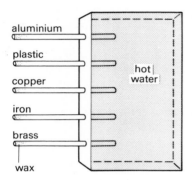

The rods are coated with wax and the trough is filled with boiling water. The wax melts first in the rod made of the best conductor, and so on. Good conductors feel cold to touch because they rapidly conduct away the heat of the hand. Bad conductors feel warm because they do not allow the heat of the hand to escape.

The mechanism of conduction is complex. In metals, free electrons are able to move and carry heat energy from high temperature to low temperature regions. Also thermal vibrations can be passed from particle to particle of the substance. In non-metals the transmission of heat energy by vibration of the particles is the important mechanism.

The safety lamp

Wire gauze is a good conductor and can stop a flame spreading.

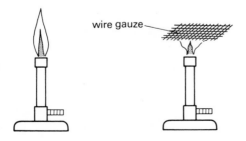

The flame does not pass through the gauze because the wire conducts the heat away and the ignition temperature of the gas is not reached.

Humphrey Davy incorporated this principle in his miner's safety lamp. The flame burns on the wick but does not ignite any explosive concentrations of methane in the mine. The miners could therefore see to work in safety.

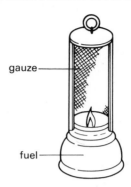

Convection

Convection currents are caused by hot liquid, which is less dense than cold liquid, rising from the bottom of the flask which is being heated. Convection currents allow heating to occur throughout a fluid by circulation.

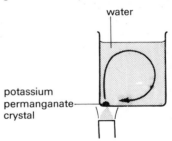

The small flame heats the large beaker of water containing a crystal of potassium permanganate. The purple colouration in the water shows how the crystal dissolves and how the solution is carried through the water by convection.

Examples of convection

a) Ventilation

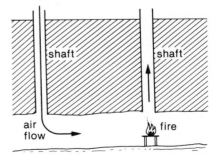

Ventilation in mines used to be done in the way shown in the diagram above.

A fire under one shaft heated the air which rose out of the mine. Fresh air was thus drawn into the mine down a neighbouring shaft.

b) The domestic hot water system

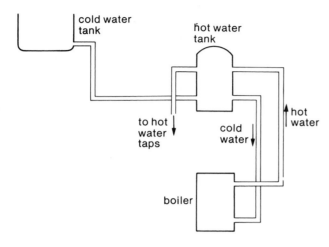

Water is heated by the boiler and it rises by convection and accumulates in the hot water tank. It rises to the top of the tank and is drawn off from there to the hot water taps. Cold water is supplied to the bottom of the hot water storage tank and boiler.

Radiation

Radiant heat consists of infrared radiation from the electromagnetic spectrum (see p. 30) and it can travel through a vacuum. When radiant heat falls on an object, some of the energy is reflected and some is absorbed. The absorbed energy heats the object and its temperature rises.

Detecting radiant heat

Radiant heat is detected using a thermopile. A thermopile is a large number of thermocouples connected together.

A thermocouple consists of wires of two different metals joined together. When the junction is heated, an electric current can be detected flowing through a galvanometer connected in a circuit as shown in the diagram below.

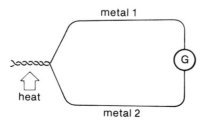

Leslie's Cube

Some surfaces are better than others at radiating heat.

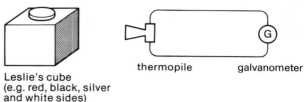

Leslie's cube
(e.g. red, black, silver and white sides)

A Leslie's cube has a different kind of surface on each face. It is filled with boiling water and placed in front of a thermopile connected to a sensitive galvanometer. The deflection of the galvanometer pointer is a measure of the heat radiating from the particular surface which is facing the thermopile.

Surfaces have different capacities for absorbing radiant energy. The best absorber is a matt black surface (poor reflector) and the worst absorber is a highly polished silver surface (good reflector).

Applications of radiation

a) Buildings painted white keep cool in summer as radiant heat is reflected.
b) White clothing reduces absorption of heat in hot climates.
c) Silver teapots keep tea warm longer as shiny surfaces are poor radiators of heat.
d) The greenhouse

Infrared radiation from the Sun has a short wavelength and it passes through glass without being absorbed. It is absorbed by the plants in the greenhouse, raising their temperature. The plants also radiate heat, but of a much longer wavelength which cannot penetrate glass. The greenhouse is able to trap heat and can do this efficiently even on a dull day.

The vacuum flask

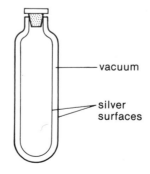

The silver surfaces on the inside of the double walls reduce the radiation of heat. The vacuum helps prevent the conduction of heat. Heat losses can occur from the neck by conduction through the glass walls and the cork but these are very small. There is little movement of heat into or out of a sealed vacuum flask so hot liquids stay hot and cold liquids stay cold inside.

Exercises

1 How can you show that heat travels more quickly through copper than iron?

2 Name the three ways in which heat is transmitted. Which is the most important in   a) solids   b) liquids?

3 Discuss how heat is lost from a bowl of soup standing on a table. How could heat losses be reduced, thus keeping the soup hot longer?

4 Why is a teapot sometimes given a shiny outer surface? How does a knitted tea-cosy help to keep the pot and its contents hot?

5 A coal fire not only heats a room but also helps to ventilate it. How does this happen?

6 Draw a labelled diagram of a vacuum flask. Show how heat losses are kept to a minimum.

7 Explain how the design of a Davy lamp helped to reduce the risk of explosions in coal mines.

8 How can you show that heat travels through air mainly by convection?

9 How can you show that water is a poor conductor of heat?

10 How does heat reach the Earth from the Sun?

6 Magnetism and electricity

6.1 Magnetism

Magnetic substances

A substance which is attracted by a magnet is called magnetic. Iron, nickel, cobalt and some alloys containing these metals are all strongly magnetic.

The poles of a magnet

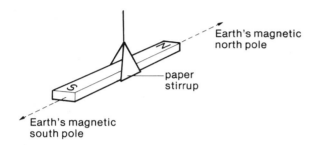

If a magnet is freely suspended, it will move until it comes to rest in a north-south direction. The end of the magnet which points towards the Earth's magnetic north pole is called the north pole of the magnet. The north pole of the magnet is therefore the 'north-seeking pole'. Similarly, the end of the magnet which points south, i.e. the 'south-seeking pole' is called the south pole of the magnet.

If the magnets are brought close together, *like* poles will *repel* each other and *unlike* poles will *attract* each other.

Testing for a magnet

If a known pole of a magnet is brought up to the material under test and attraction occurs, the test material is magnetic but may or may not be a magnet. If the opposite pole of the magnet is then brought up to the test material and repulsion occurs, the material under test is a magnet.

Induced magnetism

When a piece of unmagnetized but magnetic material is placed close to or in contact with a bar magnet it becomes magnetized by induction.

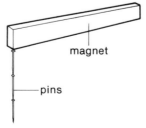

When a chain of pins is picked up by a magnet, each pin becomes an induced magnet. The poles are induced so that the pins attract one another.

The magnetic properties of iron and steel

Iron is an element and steel is an alloy of iron and carbon. When pieces of iron and steel are placed close to, or in contact with, a magnet they become induced magnets. The iron is magnetized more strongly than the steel. However, when the permanent magnet is removed the steel keeps its magnetism better than the iron.

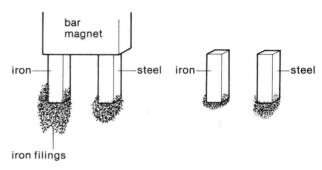

We say that iron is magnetically soft and steel is magnetically hard.

Methods of making magnets

a) Electrical method

An electric current passing through a wire has a magnetic effect which can be used to magnetize a steel bar.

The steel bar is placed inside a cylindrical coil of insulated copper wire called a solenoid. The solenoid is then connected to a 12 V battery as shown in the diagram:

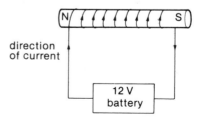

The current is switched on and then off immediately. The magnetic poles are produced as shown in the diagram and can be reversed by reversing the direction of the current.

b) Stroking methods

i) **Single touch method**

A steel bar is stroked from end to end with a magnet.

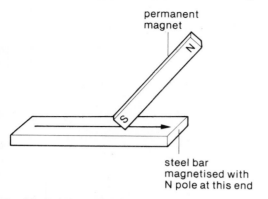

ii) **Double touch method**

A steel bar is stroked with two magnets simultaneously:

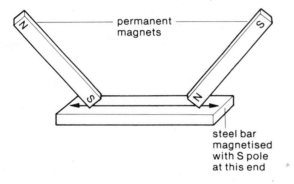

Demagnetization

A magnet that is dropped and knocked about will lose some of its magnetism. Magnets can also be demagnetized by an electrical method. The magnet is placed in a 500 turn solenoid which is facing east-west. A 12 V alternating current is supplied to the solenoid and the magnet is withdrawn slowly to a distance of several metres whilst the current is still flowing.

The alternating current disrupts the magnetism. The magnet is held in an east-west direction so that the Earth's magnetic poles have no effect on the demagnetization.

Another method of demagnetization is to alternately heat and cool the magnet. Again the magnet should be placed in an east-west position.

Magnetic fields

The space around a magnet in which a magnetic effect can be detected is called a magnetic field. The magnetic field around a bar magnet can be studied using iron filings.

A bar magnet is placed beneath a sheet of stiff white paper and a thin layer of iron filings is sprinkled on the paper. When the paper is gently tapped, the pattern of the magnetic field becomes clear.

A plotting compass can also be used to plot the magnetic field around a bar magnet. A plotting compass is a small magnet which is pivoted in the middle and enclosed in a glass case.

The bar magnet is placed on a sheet of paper and the compass is placed near one pole of the magnet.

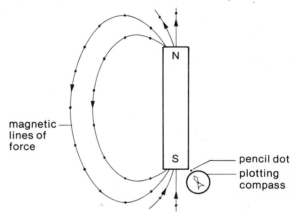

A pencil dot is made on the paper next to the point of the compass (north pole). The compass is then moved so that the tail of the compass (south pole) is next to the dot. This is repeated until the compass reaches the other pole of the bar magnet or goes off the paper. Other lines are plotted in the same way.

The accepted way of drawing magnetic fields is as a series of lines with the direction from the north pole to the south pole.

The magnetic field between two *unlike* poles looks like this:

The magnetic field between two *like* poles looks like this:

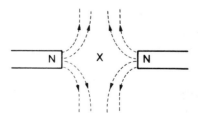

There is a neutral point at X. This is where the effects of the two fields from the bar magnets are equal and opposite. There is no resultant magnetic field at X.

The Earth's magnetism

The Earth has a magnetic field similar in shape to that of a bar magnet:

There is no actual bar magnet inside the Earth; the magnetic field is probably caused by the swirling of the Earth's molten core.

The Earth's magnetic and geographical north poles do not lie in the same place. The angle between magnetic north and geographical north is called the *declination*.

Notice that the magnetic pole to the north is in fact a south pole. This agrees with the experience that a freely suspended bar magnet comes to rest with its north pole pointing to the north.

If a compass needle is allowed to swing freely in a vertical plane, it will come to rest lying along one of the lines of the Earth's field:

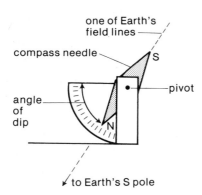

The angle between the compass needle and the horizontal is known as the angle of dip. It varies all over the world. In England it is about 70°, on the equator it is 0°, and above the magnetic north pole it is 90°.

Exercises

1 Which three elements are strongly magnetic?

2 How would you test a bar of iron with a known magnet to see if the iron is also a magnet?

3 You are given two identical iron bars and told that one of them is a magnet. How can you tell which one is the magnet without using any other equipment?

4 The diagram shows two bar magnets being stored with keepers.

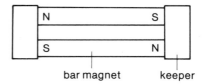

The keepers are made of iron and become induced magnets in these positions. Copy the diagram and mark on it the induced magnetic poles.

5 Describe any one method for making a magnet from an iron rod. Draw a diagram of the equipment used, showing clearly the positions of the magnetic poles on the new magnet.

6 Draw the magnetic field:
   a) around a bar magnet;
   b) between the poles of a horseshoe-shaped magnet;
   c) between two south poles.

7 Draw a diagram of the lines of the Earth's magnetic field as they pass through your laboratory. Mark clearly the angle of dip.

6.2 Electrostatics

There are many everyday examples of electrostatics:
a) Rub a plastic pen on your sleeve and then pick up small pieces of paper with it.
b) Rub a balloon on your sleeve; the balloon will stick to the wall.
c) Lightning.
d) Crackling and/or sparks when you take off a jumper.

Two kinds of charge exist: positive and negative. Rubbing a *polythene* rod with a cloth charges the rod with *negative* charge. Rubbing a *cellulose acetate* rod with a cloth gives the rod a *positive* charge.

If two polythene rods are rubbed with a cloth, they will repel each other. One rod is held and the other pivoted on top of two watch glasses:

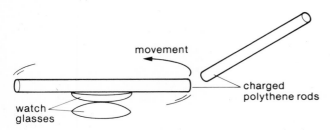

If a polythene rod, pivoted in the same way as before, is tested with a rod made of cellulose acetate instead of polythene, attraction occurs. The pivoted rod moves towards the cellulose acetate rod and not away from it. *Like* charges *repel* each other and *unlike* charges *attract* each other.

The electron theory

All matter is made up of tiny particles called atoms. Atoms themselves are made up of three basic particles called electrons, protons and neutrons.

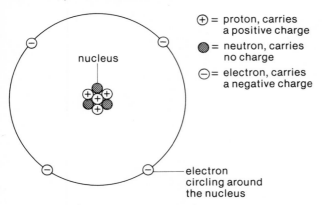

Protons and neutrons are part of the nucleus of the atom and electrons circle around the nucleus (see p. 89).

Electrons have a negative charge, protons have a positive charge and neutrons have no charge.

A material that carries no resultant charge contains equal numbers of positive and negative charges, i.e. equal numbers of protons and electrons.

It is sometimes possible for electrons to separate from an atom. The movement of electrons can cause either a positive or a negative charge on a material.

If the atoms in the material lose electrons, the material becomes positively charged and conversely if the atoms gain electrons, the material becomes negatively charged.

When a polythene rod is rubbed with a cloth, some electrons from the cloth attach themselves to the polythene so that the polythene becomes negatively charged.

When a cellulose acetate rod is rubbed, electrons move from the rod to the cloth so that the rod becomes positively charged.

Earthing

A negatively charged body and a positively charged body can both lose their charges by being connected to earth. The extra electrons on a negatively charged body flow to earth. A positively charged body gains electrons which flow from earth.

Insulators and conductors

Insulators will *not* allow electrons to flow. The electrons in the atoms of an insulator are firmly bound and will not move from atom to atom.

Conductors allow electrons to flow. The electrons are loosely bound to atoms and move freely from one atom to another.

Polythene is an insulator. When a polythene rod is rubbed with a cloth, electrons collect on the surface of the rod. The electrons do not flow away to earth through the person who is holding the rod, as they would if the rod were a conductor.

The gold-leaf electroscope can be used to detect charge:

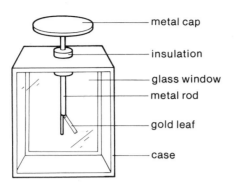

The electroscope consists of a metal cap and a rod. The lower end of the rod is flattened and the rod is enclosed within a glass case but insulated from the case. There is a thin leaf of gold attached by one edge to the metal plate.

If a negatively charged body is brought close to, but not touching, the cap of the electroscope, the gold leaf moves away from the plate. This is because the negative charge repels electrons from the cap and they travel down into the leaf and plate. The leaf and plate are therefore both negatively charged and so they repel each other. When the negatively charged object is removed, the leaf collapses back again.

If a positively charged body is brought close to the cap, the gold leaf also moves away from the plate but this time electrons are attracted to the cap: the plate and the leaf both become positively charged.

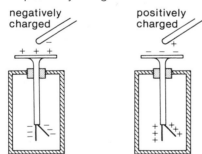

Charging an electroscope by electrostatic induction

The gold leaf in the diagram above is charged even though a charged body has not touched it. It is said to have been charged by *induction*. When the charged rod is removed, however, the electrons redistribute themselves so that the electroscope no longer has a separation of charge. If the cap of the electroscope is earthed whilst the rod is present, the electroscope keeps a charge.

a) To give an electroscope a positive charge:

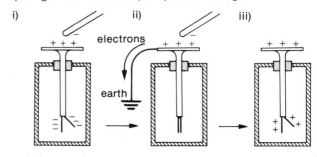

 i) Bring a charged polythene rod near the electroscope so that there is an induced charge on the leaf.
 ii) Momentarily earth the cap, allowing the electrons on the leaf to escape to earth.
 iii) Remove the earth and then the rod, and the electroscope is left with a positive charge.
b) To give an electroscope a negative charge: Repeat a) above but use a charged cellulose acetate rod.

Testing for charge sign

If a negatively charged body is brought close to the cap of a negatively charged electroscope, electrons are repelled from the cap and the leaf diverges further from the plate.

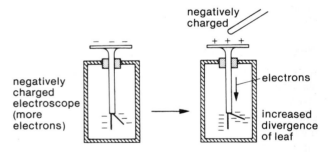

If a positively charged body is brought close to the cap, electrons are attracted upwards towards the cap and the leaf collapses to the plate.

When an uncharged metal bar is brought close to the cap of a charged electroscope, the leaf collapses. This is because induction takes place in the bar. Electrons are attracted to or repelled from the end of the bar and this leads to a rearrangement of electrons in the electroscope.

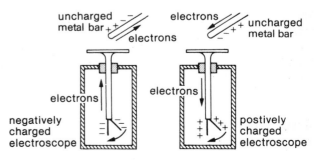

The lightning conductor

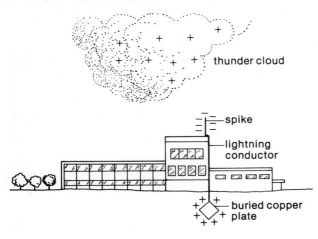

The thunder cloud passing over the school building induces a charge separation in the lightning conductor.

The electrons gathering around the spike are so numerous that they repel any negatively charged air particles which move upwards towards the cloud. This lessens the chance of a lightning strike on the building.

Electrons from earth neutralize the positive charge on the buried plate and this helps to increase the discouraging charge on the spike.

Charge and potential difference

A charged body has either an excess or a deficit of electrons. Charge is measured in coulombs (C) and each electron carries a negative charge of about 1.6×10^{-19}C. Charge moves through a conductor from one place to another when there is a potential difference (p.d.) between the two places. When a negatively charged conductor is connected to earth there is a flow of electrons from the conductor to earth because of the potential difference between the conductor and earth. Earth is at the zero of potential.

If a solid metal ball on a polythene stand is given a charge, the charge will distribute itself around the surface of the ball until there is no potential difference between any two places on the ball.

The divergence of a gold leaf in an electroscope shows that there is a potential difference between the cap and the case: if the two are joined by a conducting wire there will be a flow of charge and the leaf will collapse. Potential difference is measured in volts (V) and is discussed again on page 72.

Capacitance

The potential of a conductor increases when the charge on it increases. The size of the increase in potential depends upon the size of the conductor: the larger the conductor, the smaller the increase in potential. The ratio of the charge (symbol Q) on the conductor and its potential V is known as the capacitance C of the conductor.

$$C = \frac{Q}{V}$$

The unit of capacitance is the farad (F). If the addition of one coulomb of charge to a conductor changed its potential by one volt, then the conductor would have a capacitance of one farad. In practice, capacitance is measured in fractions of a farad (e.g. 1 μF = 1×10^{-6}F).

Parallel-plate capacitor

Holding a metal plate over a gold leaf electroscope makes the leaf go down.

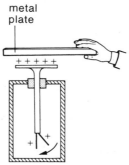

There is an induced charge separation in the metal plate and electrons are drawn up from earth into the plate.

The leaf goes down because the potential of the electroscope has dropped even though the charge on the electroscope remains the same.

Consequently, the capacitance of the electroscope has been increased.

This principle is used in the design of a simple practical capacitor, the parallel plate capacitor. One version consists of two long strips of foil separated by an insulator rolled up into a cylinder. Another, shown below, is used in the tuning of radios. Turning the shaft alters the area of overlap of the plates and the capacitance of the capacitor.

Exercises

1. What are the two kinds of charge and how do you obtain them?
2. Draw a labelled diagram of a gold leaf electroscope.
3. Explain how a gold leaf electroscope is given a negative charge. How could the charged electroscope be used to detect a positive charge?
4. The girl in the photo is holding onto an electrostatic generator and has become charged.

Why is her hair standing up like that?

5. A thin stream of tap water is attracted to a charged rod, no matter what charge is on the rod. How can this be explained using the idea of electrostatic induction?
6. A large ball and a small ball, made of the same metal, are given an equal charge. What can you say about the potential and capacitance of the large ball compared to the small one? Which way would charge flow if the two were joined by a wire?
7. A metal plate is held over a charged electroscope and the leaf drops. What has happened to the potential and capacitance of the electroscope? What would be the effect of lowering the plate nearer to the cap?
8. What is the unit of capacitance? What is the capacitance of a conductor if it has a potential of 250 V and carries a charge of 2.5×10^{-5} C?

6.3 Current and resistance

An electric current is a movement of electric charge. By convention, the current is said to flow in the direction of movement of positive charge. However, electrons are the particles that actually move and they carry the negative charge.

Current is measured in amperes (A). One *ampere* of current flows around a circuit if one *coulomb* of charge passes around the circuit in one *second*.

The charge on an individual electron is -1.6×10^{-19} C, so a current of one ampere means something like 6.2×10^{18} electrons are moving through the conductor in one second.

Potential difference (p.d.)

A current will only flow if there is a complete circuit to flow around. The current can only get *through* a conductor in the circuit if there is a potential difference *across* the conductor. As the current moves through the conductor, it gives up some of its energy to the conductor. The greater the potential difference, the greater is the amount of energy given up. A potential difference of one *volt* means that one *joule* of energy is given up by each coulomb of charge passing through the conductor.

Electron moving force (e.m.f.)

There may be many component parts to one circuit, each requiring a potential difference across its connections so that the current flows. The sum of all of the individual potential differences in a circuit is the electron moving force (sometimes called electromotive force). The term e.m.f. should only be applied to the source of the electrical energy such as a power pack or a cell. We say for example, that a battery has an e.m.f. of 3 volts.

Resistance

The relationship between potential difference and current for a conductor, such as a wire, has been investigated and is known as **Ohm's Law**: *the current passing through a wire is proportional to the potential difference across its ends, provided the temperature remains constant.*

The law can be expressed as

$$\frac{\text{potential difference}}{\text{current}} = \text{a constant}$$

The constant is known as the *resistance* of the wire and, as its name suggests, it is a measure of how difficult it is for a current to flow through the wire.

Ohm's Law is often written as:

$$\frac{V}{I} = R \quad \text{or} \quad V = IR \quad \text{or} \quad I = \frac{V}{R}$$

where V is the potential difference measured in volts (V), I is the current measured in amps (A), and R is the resistance measured in ohms (Ω).

The resistance of a wire depends on
a) the material that it is made from,
b) its dimensions.

In general, for a wire of any named material:
i) the resistance increases as the length increases;
ii) the resistance increases as the cross-sectional area decreases.

The current flowing in a circuit may be increased or decreased by altering the resistance of the circuit.

A variable resistance is called a *rheostat*. Standard resistances for inclusion in radios etc. are known as resistors. Here are the circuit symbols for a resistor and a rheostat:

resistor   rheostat

Verification of Ohm's Law

Connect up the following circuit:

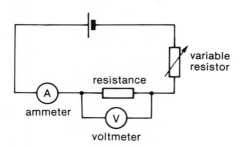

Note that current is measured with an ammeter placed in *series* with the resistance. Potential difference is measured with a voltmeter in *parallel* with the resistance.
a) Close the switch and take the ammeter and voltmeter readings.
b) Alter the variable resistance to give a range of current values. Take ammeter and voltmeter readings after each alteration.
c) Plot a graph of potential difference V against current I.

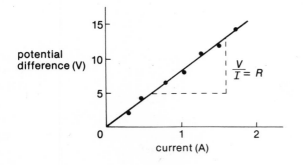

A straight-line relationship verifies Ohm's Law; the gradient of the line is the resistance in ohms.

If the temperature of the resistor is not kept constant, the graph will not be a straight-line relationship. For example, a graph similar to the one below is obtained if the experiment is repeated with a torch bulb (filament lamp) in place of the fixed resistance.

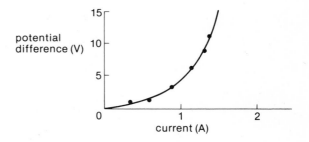

The resistance (gradient) increases as the current through the filament increases, and therefore the temperature of the filament goes up.

Series and parallel

In a series circuit, the current only has one path to take but in a parallel circuit, it has more than one path.

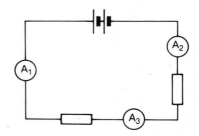

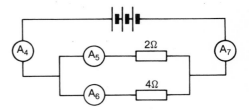

Consequently:

A_1, A_2 and A_3 read the same;
A_4 and A_7 read the same;
A_5 reads twice as much as A_6;
$A_5 + A_6 = A_4$ (or A_7)

Ammeters have a very low resistance and do not appreciably alter the current.

In any circuit, the sum of the individual potential differences is equal to the e.m.f. of the source of electricity.

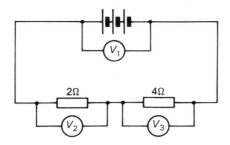

$V_1 + V_2 + V_3$ = e.m.f. of the battery

The potential difference across the connecting wires is very small because the connecting wires have such a low resistance.

V_1 registers a reading because the battery has an *internal resistance* and there is current through it.

Resistances connected in parallel have the same potential difference across them.

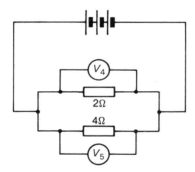

Consequently, V_4 and V_5 give the same reading.

Voltmeters have a very high resistance and do not draw off an appreciable amount of current.

Resistors in series

Three resistors R_1, R_2 and R_3 are connected in series and the potential difference across each is measured.

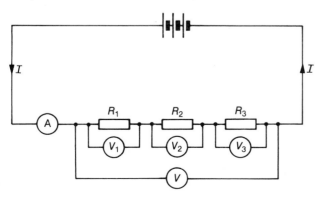

The current I passes through each resistor in turn and the overall potential difference V is the sum of the individual potential differences.

$V = V_1 + V_2 + V_3$

Using Ohm's Law, $V = IR$, we can write

$IR = IR_1 + IR_2 + IR_3$

where R is the total resistance of the resistors together.

Dividing both sides by I gives

$R = R_1 + R_2 + R_3$

So the total resistance of a number of individual resistances joined in series is simply the sum of those resistances.

Resistors in parallel

If the resistors are connected now in parallel, each will take a different current.

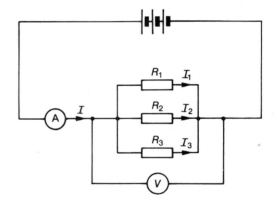

There is a common potential difference across each resistor and the current I is the sum of the current passing through the individual resistors.

$I = I_1 + I_2 + I_3$

Using Ohm's Law in the form $I = \dfrac{V}{R}$ we can write

$\dfrac{V}{R} = \dfrac{V}{R_1} + \dfrac{V}{R_2} + \dfrac{V}{R_3}$

where R is the total resistance of the resistors together.

Dividing both sides by V gives

$\dfrac{1}{R} = \dfrac{1}{R_1} + \dfrac{1}{R_2} + \dfrac{1}{R_3}$

So in this case the addition is not a simple sum. Notice that joining resistors in parallel *lowers* the resistance because you are providing an additional route for the current.

Exercises

1 Copy and complete these sentences.
One _____ of current flows around a circuit if one _____ of charge passes around the circuit in one _____ .

A potential difference of one _____ shows that one _____ of energy is given up by each _____ of charge passing through the conductor.

2 Explain the difference between potential difference (p.d.) and electron moving force (e.m.f.).

In the following circuit, the readings V_1 and V_2 added up to 4 V but the battery had an e.m.f. of 4.5 V. Explain the difference.

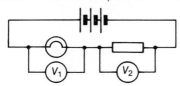

3 State Ohm's Law and give three different ways of writing it in symbols. What are the units of potential difference, current and resistance?

4 Here is a table of results taken from a typical Ohm's Law experiment:

potential difference (V)	current (A)
2.1	0.2
4.6	0.4
6.0	0.6
8.1	0.8
12.5	1.2

Plot a graph of p.d. against current and find the resistance.

5 Here is a table of results from an ammeter/voltmeter verification of Ohm's Law on a filament lamp taken from a torch.

potential difference (V)	current (A)
0.5	0.1
1.5	0.2
2.8	0.3
4.3	0.4
7.2	0.5

Comment on the shape of the graph of p.d. against current and find the resistance of the bulb when taking a current of 0.35 A.

6 How many different ways can three 5Ω resistors be connected?

7 Calculate the effective resistance of three 10Ω resistors connected a) in series b) in parallel.

8 What is the potential difference across a 15Ω resistor if a current of 3 A is flowing through it?

9 A battery can supply an e.m.f. of 12 V and has an internal resistance of 2 Ω. The circuit it is connected to consists of one 4 Ω resistor and a 2 Ω resistor in series. What is the total resistance of the circuit? What is the p.d. across the 4 Ω resistor?

10 Study this circuit diagram:

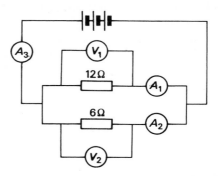

If the battery has no internal resistance, what is the resistance of the circuit?
A_1 reads 1 A. What do A_2 and A_3 read?
What do V_1 and V_2 read?

6.4 Electrical energy

Electricity is a form of energy measured in joules. It can be converted into work as it passes through a potential difference. The work done by one coulomb of electricity passing through a p.d. of one volt is one joule. If this work is done in one second then we can say that the work done by one ampere of electricity passing through one volt is one joule.

Summarizing:

work done = p.d. × current × time
(joules)   (volts) (amps)   (seconds)

The *power* of an electrical appliance is a measure of the rate at which that appliance turns electricity into another form of energy. For example, an electric light bulb turns electric energy into light, an electric fire turns electric energy into heat, and an electric motor turns electric energy into mechanical energy.

Power is measured in watts (W). If an appliance is working with a power of 1 W, it is turning 1 J of electricity into another form of energy in 1 second.

$$\text{power (in watts)} = \frac{\text{work (joules)}}{\text{time (seconds)}}$$

$$= \frac{V \times I \times t}{t} = V \times I$$

where V = p.d. (volts)
I = current (amps)
t = time (seconds)

Example 1

Find the power of an electric lamp operating on a 240 V supply using a current of 0.25 A.

power = p.d. × current
      = 240 × 0.25
      = 60 W

Example 2

Find the current used by an electric fire rated at 2000 W operating on a 240 V supply.

current = $\frac{\text{power}}{\text{p.d.}}$

= $\frac{2000}{240}$

= 8.3 A

The kilowatt hour

1000 W = 1 kilowatt (kW)

Thus an appliance with a power of 1 kilowatt will turn 1000 J of electricity into another form of energy in 1 second. If, for example, an electric fire with a power of 1 kW burns for 1 hour it will turn 3 600 000 J of electricity into heat (1000 × 60 × 60 = 3 600 000).

When electricity bills are calculated, the unit of energy used is the kilowatt hour (kWh). 1 kWh is the amount of energy that is used by an appliance working at the rate of 1000 W for 1 hour.

Example 3

How many units of electricity will a 2 kW fire use if it is turned on for 3 hours?

total units used = 2 × 3
                = 6 kWh

Example 4

How many units are consumed by a 750 W hairdryer blowing for 15 minutes?

total units used = 0.75 × 0.25
                = 0.1875 kWh

Example 5

How much will it cost to run a 2000 W fire for 30 minutes if electricity costs 6p a unit?

units consumed = 2 × 0.5 kWh
              = 1 kWh
cost = 6p

Heating and resistance

The heating element in an electric fire consists of a coiled resistance wire which becomes very hot when a current is passed through it. The work done by the current becomes kinetic energy in the vibrations of the molecules of the wire.

The work done is related to the current.

work done = $V \times I \times t$

This expression can be combined with Ohm's Law for the resistance

$V = I \times R$

work done = $I^2 \times R \times t$

This work becomes heat energy in the wire so heat produced = $I^2 \times R \times t$ joules.

where I = current (A)
R = resistance (Ω)
t = time (s)

Example 6

How much heat is produced when a wire of resistance 100 Ω carries a current of 5 A for 30 minutes?

30 minutes = 30 × 60 seconds
heat produced = I^2Rt
  = $5^2 × 100 × 1800$
  = 4 500 000 J

Electricity in the home

The amount of electricity required for domestic and industrial use varies from one time to another. To allow for sudden changes in demand, power stations in Britain are linked by cables. This allows the electrical energy produced in one place to be used in another. This system is called the National Grid.

The mains electricity is supplied to each house via a service cable. For electricity to flow, a complete circuit is needed so the service cable therefore contains two wires. One of these carries the current into the house and is called the 'live' wire and the other wire is the return or 'neutral' wire. The supply to the house has a p.d. of 240 V.

The electricity is supplied in the form of an alternating current. The current changes direction 50 times a second and is therefore said to have a frequency of 50 hertz (Hz) (see p. 84).

The voltages used in the National Grid are very high and the current is relatively low so that less energy is lost as heat. When the electricity reaches its destination the voltage is reduced by a transformer (see p. 84).

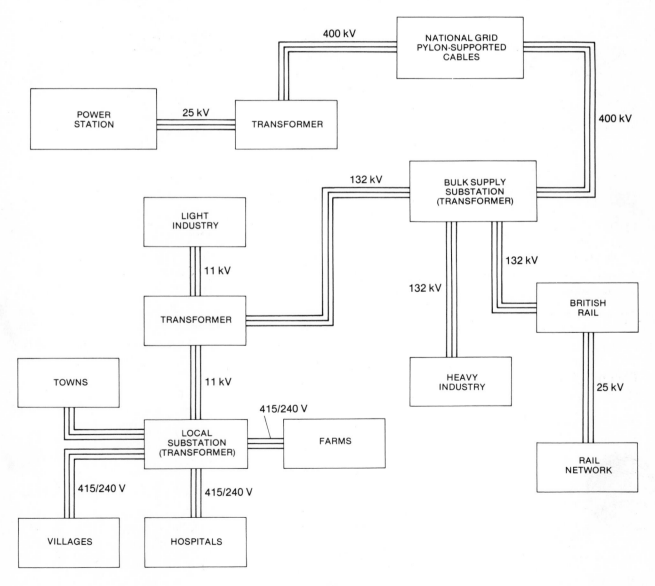

The circuit inside the house usually consists of one or more complete loops of cable to which outlet sockets are connected.

As well as live and neutral wires, power circuits in a house have a third wire which is connected to earth. The wire is used for earthing the metal cases of any appliances used. It is a safety measure to prevent anyone from receiving a shock should the casing become live. If any part of the appliance carrying current accidentally touches the casing of the appliance, the current flows through the earth wire to earth and not through anyone who happens to touch the casing. The earth wire has a low resistance and the current flow will be large enough to 'blow' the fuse.

Fuses

A fuse is a most important safety device: it is a weakness deliberately built into an electrical circuit. It consists of a short length of thin wire. When the current in the circuit reaches a certain value, the fuse melts and the circuit is broken. Without a fuse, some other length of wire in the circuit could become hot and start a fire. Each fuse is designed so that the maximum current it can take before melting is just greater than the normal safe working current for the circuit.

Every household circuit is protected by a fuse near where the mains supply enters the house. There are separate fuses for each circuit in the house. The lighting, water heating and cooker have their own circuits as they require fuses of differing sizes.

The plugs which connect appliances to mains sockets usually contain a cartridge fuse.

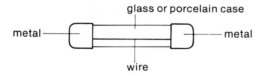

They can be rated at 3 A (blue) or 13 A (brown). Fuses of 3 A are used for appliances with a power of up to 720 W; fuses of 13 A are used for all appliances having a power greater than 720 W.

Wiring a plug

An electrical appliance is connected to a mains supply socket via a lead and a three-pin plug. The lead contains three insulated wires. The insulation on the live wire is coloured brown and this wire is connected through a fuse to the pin marked L (live).

The correct fuse for the particular appliance should always be used otherwise the appliance and possibly other parts of the circuit may be damaged by an excess current.

The insulation on the neutral wire is coloured blue and this wire is connected to the pin marked N (neutral).

The earth wire is green and yellow and is connected to the pin marked E. The earth wire connects the casing of the appliance to the earth.

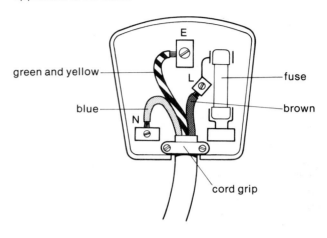

Exercises

1 How many joules of energy does a 1 kW fire give out a) in one second b) in one minute c) in one hour?

2 What is the power of an electric light bulb which draws a current of 0.41 A from a 240 V supply?

3 What is a kilowatt hour? What does it measure? How many units of electricity will a 2 kW fire use if it is switched on for a) one hour b) five hours c) five minutes?

4 How much heat is given off from a wire of resistance 200 Ω carrying a current of 2 A for ten minutes?

5 Electricity needs a complete circuit before it flows. So electricity comes in to the house from the power station, runs through the appliances and then goes back through another wire to the power station. If the power station gets all of its electricity back, why should we pay electricity bills?

6 Explain why fuses are needed. Where would you expect to find fuses in your house?

7 What size of fuse would you recommend for an electric drill of power 315 W to be run from the normal mains supply (240 V)?

8 A boy fitted a 3 A fuse to a plug connected to an iron (rated 850 W) and plugged it into the mains (240 V). What happened and why?

9 Draw a labelled diagram of a fused three-pin plug. Explain why it is important to connect the wires correctly.

10 Explain why the earth wire is especially important for appliances having a metal case. Where is the earth wire connected?

6.5 Electromagnetism

A magnetic field can be detected around a straight length of wire when it is carrying a current. The field may be plotted using iron filings or a small compass (p. 67).

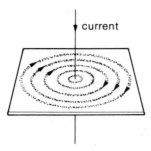

The right-hand grip rule

The direction of the magnetic lines of force can be worked out using this rule. Imagine that the wire is gripped in the right hand and that the thumb is pointing along the wire in the direction of the current. The fingers then curl around the wire in the direction of the line of force.

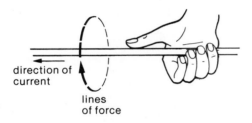

The magnetic field around a solenoid carrying a current is similar to that of a bar magnet.

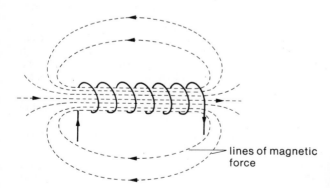

Notice that the field inside the solenoid is made up of straight parallel lines of force.

If a current is passed through a circular coil the following pattern is obtained.

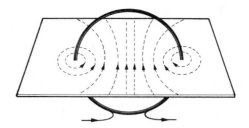

The poles of a solenoid field

Look at the coil from one end. If the current is flowing in an anticlockwise direction, then that end is the north pole. Alternatively, if the current is flowing in a clockwise direction, then that end is the south pole.

Electromagnets

When a piece of iron is placed inside a solenoid which is carrying current, the iron becomes strongly magnetized. When the current is switched off, the iron ceases to be a magnet. This arrangement is called an electromagnet. The solenoid alone forms a magnetic field as already described, but this is only weak. Ferrous scrap (iron) and non-ferrous scrap can be sorted out using an electromagnet.

The large magnet is switched on as it is brought up close to a pile of scrap. The ferrous scrap is attracted to the magnet, which is then lifted up with the scrap clinging to it. The magnet can then be moved away and the ferrous scrap dropped somewhere else when the magnet is switched off.

The electric bell

The electric bell contains a U-shaped soft iron core which has two solenoids wound in opposite directions around the two arms.

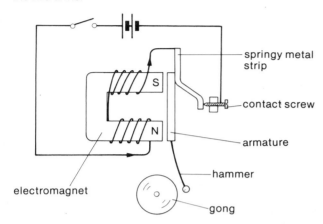

The electromagnet attracts a piece of iron called the armature. As the armature moves towards the electromagnet the contacts separate and so the circuit is broken. The electromagnet is therefore no longer magnetized and so the armature springs back and the contact is renewed. This process is rapidly repeated causing the hammer to strike the gong each time.

The magnetic relay

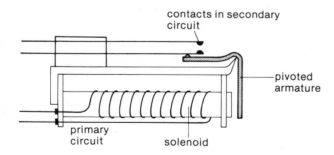

The magnetic relay uses a small current in a primary circuit to control a larger current in a secondary circuit. When a burst of electricity passes through the solenoid, it creates a magnetic field which attracts the armature. This pushes the contacts together and allows current to flow in the upper circuit. When electricity ceases to flow in the solenoid the contacts spring apart.

Magnetic relays are used in telephone circuits to operate switches that select the required telephone number.

The motor effect

When a current flows through a wire which is in a magnetic field, a force acts on the wire. This is called the motor effect and it can be explained by thinking about the interaction of the two fields.

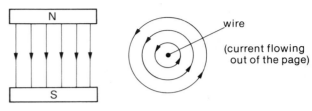

The circular field of the current-carrying wire is superimposed on the uniform field on the left. In one area the lines of force pull together, in another area they pull against each other to produce a force in one direction.

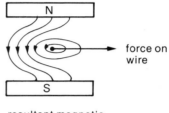

resultant magnetic field

The direction in which the wire will move can be predicted using *Fleming's left-hand rule:* if the thumb, first finger and second finger of the left hand are held at right angles to each other

a) the thumb points in the direction of the **motion** of the wire;
b) the first finger points in the direction of the lines of **force** of the magnet;
c) the second finger points in the direction of the **current**.

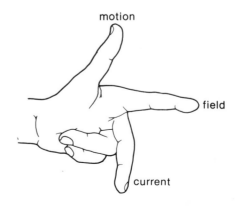

Moving-coil galvanometer

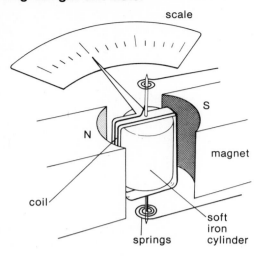

This device is used to detect and measure very small currents (of the order of milliamps). The current enters the coil through control springs. The coil rotates because of the motor effect until prevented by the springs. The greater the current, the greater is the deflection. The soft iron cylinder is stationary and, along with the shaped magnetic pole pieces, produces a *radial* magnetic field:

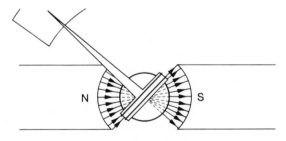

The radial field ensures that the pointer moves an equal amount for an equal increase in current, i.e. the scale is *linear*.

Ammeter

The moving-coil galvanometer can be altered so that it can be used to measure a large current. A *shunt*, which is a low resistance, is connected in parallel with the galvanometer. Most of the current to be measured passes through the shunt and only a known fraction of the current goes through the galvanometer. The value of the shunt determines the range of current that the ammeter can measure.

The resistance of the ammeter as a whole is small so an ammeter can be placed in series in circuits: it does not appreciably alter the current.

Voltmeter

A moving-coil galvanometer is converted to a voltmeter by connecting a resistance (a *multiplier*) of high value in series with the galvanometer. This cuts down the current so protecting the galvanometer. The value of the multiplier fixes the range of potential differences that the voltmeter can measure.

The resistance of a voltmeter is high. The voltmeter is placed in parallel in a circuit so that it does not draw off an appreciable amount of the current.

The d.c. electric motor

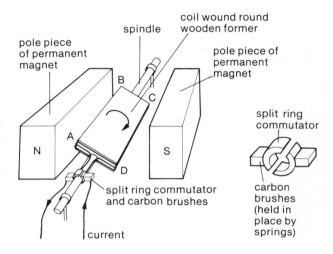

The motor likewise consists of a coil of wire through which a current can flow, suspended so that it is free to turn in a magnetic field.

The coil is mounted on a spindle so that it can rotate between the poles of a permanent magnet. The ends of the coil are connected to a split ring (a commutator). The carbon brushes press lightly against opposite sides of the commutator. The brushes are connected to the battery.

How the motor works

The motor shown in the diagram above is connected to a d.c. supply of electricity such as a battery. If current is flowing and the coil is in a horizontal position, AB will experience a force pushing it upwards and CD will experience a force pushing it downwards. This can be confirmed using Fleming's left-hand rule. When the coil reaches the vertical position, the brushes are opposite the gaps in the commutator and therefore no current can flow.

However, the momentum of the coil will carry it past the vertical position so that the brushes will then make contact with the opposite sides of the commutator. The current flowing through the coil is therefore reversed. Because of this, AB now experiences a downward force and CD an upward force. The forces on AB and CD rotate the coil in a clockwise direction as long as the current is flowing.

The moving-coil loudspeaker

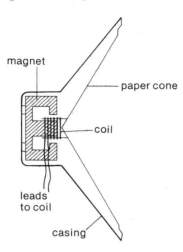

A coil, attached to a paper cone, sits in the field of a permanent magnet. When a varying electric current enters the coil, it produces a magnetic field around the coil that changes as the current changes. This magnetic field interacts with the field of the permanent magnet, causing the case to vibrate and produce sound.

Exercises

1 Draw the magnetic field you would expect to detect around a) a solenoid b) a single wire carrying a current. Mark in the directions of the fields. How does a solenoid carrying a direct current magnetize an iron bar?

2 Draw a diagram for an electric bell and explain how it works. Mark in the poles on the electromagnet of the bell and explain how you work out the polarity.

3 State Fleming's left-hand rule. When is it used?

4 Explain how you would alter a moving-coil galvanometer to make a) an ammeter and b) a voltmeter. Why is it necessary to make such alterations?

5 What changes could be made to a moving-coil galvanometer to make the pointer move further for a given current input?

6 Draw a diagram of a d.c. electric motor. Why is it necessary to have a commutator and brush arrangement?

7 Explain, with help from a diagram, how a moving coil loudspeaker speaks loudly.

6.6 Electromagnetic induction

When a bar magnet is moved into or out of a solenoid, an e.m.f. is induced in the solenoid causing a current to flow around a circuit. This is known as electromagnetic induction.

The centre-zero galvanometer shows when a current flows and gives its direction.

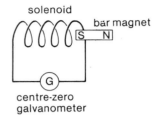

A current is registered only when the magnet is moving relative to the solenoid.

Faraday's Law

Faraday found that the size of the induced current could be increased by increasing
a) the number of turns on the solenoid;
b) the strength of the magnet or;
c) the speed of movement of the magnet.

Thus, the size of the induced current depends upon the rate at which the magnetic lines of force cut through the solenoid.

Lenz's Law

The direction of the flow of induced current depends upon the direction of the movement of the magnet.

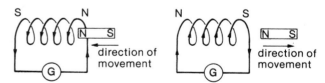

As the south pole moves into the solenoid, the current flows so that it generates a south pole to repel the magnet. On the other hand, as the south pole of the magnet is withdrawn, the current direction changes to produce a north pole to attract the magnet. The reverse polarity is obtained if the magnet is turned around.

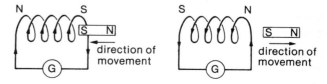

A physicist named Lenz summarized these findings into Lenz's Law:

The direction of the induced current is always such as to oppose the change producing it.

Flemings right-hand rule

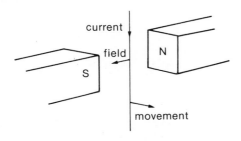

A current is induced in a straight wire if it is moved between the pole faces of two strong magnets. Fleming's right-hand rule can be used to predict the direction of such an induced current: if the thumb, first finger and second finger of the right hand are held at right angles to each other
a) the thu**m**b points in the direction of the **m**otion of the wire.
b) the **f**irst finger points in the direction of the magnetic **f**ield.
c) the se**c**ond finger points in the direction of the **c**urrent.

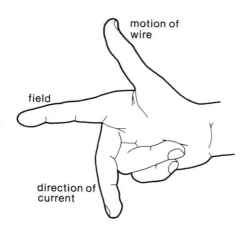

The alternating current generator

A rectangular-shaped coil rotates between the poles of a permanent magnet, called the field magnet. The ends of the coil are connected to slip rings against which carbon brushes press lightly.

As the coil rotates, it cuts across the magnetic field and so an e.m.f. is induced.
a) When the coil is vertical with side P uppermost, the sides of the coil are in the same plane as the magnetic

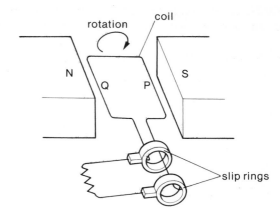

lines of force. No e.m.f. is therefore induced because the magnetic lines of force are not being cut by the coil.
b) If the coil then continues rotating, it will start to cut through the magnetic lines of force and an e.m.f. will be induced. When the coil is horizontal, the induced e.m.f. is at its maximum (the peak value).
c) As P continues on downwards, the induced e.m.f. decreases until it is again zero when the coil is vertical again but with Q uppermost.
d) The same pattern is repeated in the second half of the rotation except that the e.m.f. is reversed because the direction of movement of P and Q is also reversed.

Thus the direction of the e.m.f. induced in the coil is reversed during one complete rotation of the coil.

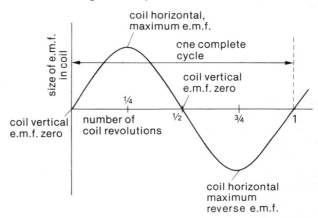

If the generator is connected in a circuit, an alternating current (a.c.) will flow. The generator is sometimes known as the a.c. dynamo.

The peak current that is delivered at the maximum e.m.f. shown is not a measure of the total effect of an alternating current. A better measure of this effect is the root mean square (r.m.s.) current which is equivalent to a direct current of about 71 per cent of the peak value of the current.

r.m.s. current = 71% of peak current

The direct current generator

The d.c. generator or dynamo is similar to the a.c. dynamo in construction except that the slip rings are replaced by a split ring commutator.

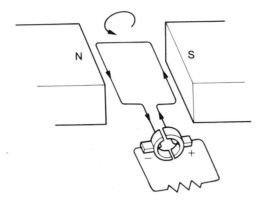

Compare the generator above with the motor on page 81 and you will see that they are identical.

The commutator is designed and positioned so that as the coil rotates, one of the brushes remains positive and the other negative. The changeover occurs when the coil is vertical, when no e.m.f. is induced. The maximum e.m.f. is generated when the coil is horizontal.

The graph shows the e.m.f. in relation to the position of the coil.

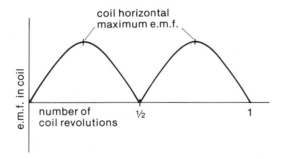

The transformer

The transformer consists of two coils of wire wound around an iron core.

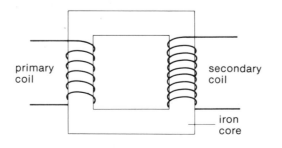

An alternating current flowing in the primary coil causes the coil to be surrounded by a continuously changing magnetic field. This magnetic field cuts through the secondary coil inducing an equally alternating e.m.f. in the coil. If the secondary coil is connected to a circuit, an alternating current will flow in that circuit.

The size of the e.m.f. in the secondary coil depends upon the number of turns of wire in the coil compared to the number of turns in the primary coil and on the size of the primary e.m.f. The number of turns is important as a consequence of Faraday's Law (p. 82).

In summary

$$\frac{\text{primary e.m.f.}}{\text{secondary e.m.f.}} = \frac{\text{number of turns in primary coil}}{\text{number of turns in secondary coil}}$$

When the primary e.m.f. is smaller than the secondary e.m.f., the transformer is known as a step-up transformer.

A step-up transformer

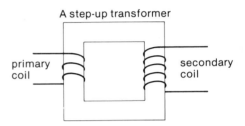

When the primary e.m.f. is larger than the secondary e.m.f., the transformer is known as a step-down transformer.

A step-down transformer

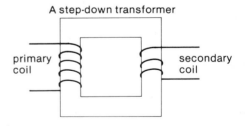

The total amount of energy coming out of a transformer cannot exceed the total amount of energy going into it. In practice, we always get less energy out of a transformer than we put in.

In an ideal transformer:

energy input = energy output

Or, if we think of the exchange per second:

power input = power output

i.e. (p.d. × current) in the primary =
(p.d. × current) in the secondary.

The efficiency of a transformer is the ratio of the usable energy given out by a transformer to the energy put into it.

efficiency (as a percentage)
$$= \frac{\text{energy output}}{\text{energy input}} \times 100$$
$$= \frac{(\text{p.d.} \times \text{current}) \text{ in the secondary}}{(\text{p.d.} \times \text{current}) \text{ in the primary}} \times 100$$

Transformers in use

Transformers play an important part in the transmission of electrical energy from the power station to the consumer.

Often the electrical energy has to be transmitted for long distances before it reaches the consumer. If the electricity were to be sent with a high current, there would be a large heating effect and a lot of energy would be lost. Energy losses are therefore minimized by using a high voltage and making the current as low as possible.

The National Grid voltage of up to 400 000 volts has to be reduced to 240 volts for use in the home; transformers are used for this. It is important that the transformer works efficiently so that large energy losses do not occur. Some ways of reducing the energy loss are as follows.
a) Use low resistance copper wire for the coils so that the heating effect is reduced.
b) Use soft iron for the core as this requires only a small amount of energy to magnetize and demagnetize the core, again reducing the heat generated in the core.
c) Use a laminated core which consists of a number of thin sheets of iron with layers of insulation glued between them. This reduces the currents (known as eddy currents) which are induced in the core by the continuously changing magnetic field.

Inductors

An inductor is a coil of many turns of wire made from a good conductor and wound onto a soft iron core. When an alternating current is passed through the inductor, an alternating field is created which cuts through the inductor. This generates another e.m.f. in the inductor which, by Lenz's law, opposes the original e.m.f. The opposing e.m.f. is known as the back e.m.f.

Exercises

1 What is electromagnetic induction?

2 What did Faraday discover about the size of the induced current and how it depends upon the coil and magnet used?

3 State Lenz's Law.
In the diagram below the magnet is dropped through the coil. As it does so currents are induced in the coil which generate magnetic fields around the coil. Which pole (N or S) is produced at X as the magnet enters the coil? Which pole is produced at Y as the magnet leaves the coil?

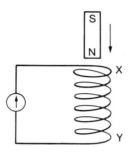

4 When is Fleming's right-hand rule used? When is Fleming's left-hand rule used?

5 Draw a diagram of a simple a.c. generator. Then draw a graph showing a typical alternating e.m.f. and mark on the graph the positions of the coil corresponding to the maximum (positive and negative) of the e.m.f. and when the e.m.f. is zero.

6 What is a transformer for?
Give an equation that relates the primary e.m.f. and secondary e.m.f. to the number of turns on the primary and on the secondary coils.

7 Explain how and where eddy currents occur. How are they reduced in a transformer?

8 A transformer is to be used to run a 12 V motor racing toy from the 240 V mains supply. If there are 2000 turns of wire on the primary coil how many turns must there be on the secondary coil?

9 A calculator adaptor containing a transformer is plugged into the 240 V mains supply and it delivers a 6 V, 250 mA supply to the calculator. If the adaptor converts all of the input energy into output energy, what current is drawn from the mains? If the adaptor only converts three-quarters of the input energy into output energy and still maintains a 6 V 250 mA output, what current will be drawn from the mains?

10 Discuss the importance of the transformer to the national distribution of electricity.

7 Physics of the atom

7.1 The electron

All matter is made up of tiny particles called *atoms*. The atoms of any one element are all identical and different from those of any other element.

Atoms all have at least one negatively charged *electron* that moves around a positively charged *nucleus*.

Thermionic emission

When a metal is heated, some of its electrons can get enough energy to escape from the surface. The greater the temperature to which the metal is heated, the greater is the emission of electrons. The electrons do not go far, however, because of the electrostatic attraction of the nuclei in the metal surface; they gather around the surface providing what is known as the *space charge*.

Cathode rays

The electrons freed by thermionic emission can be persuaded to leave the metal altogether.

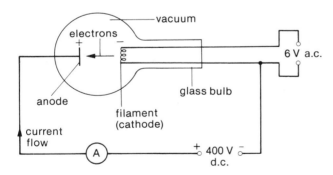

The 6 V supply heats up the filament (also known as the cathode) to provide the thermionic emission. The 400 V supply is connected so as to make the anode positive. The freed electrons are attracted across the gap between the cathode and the anode and they constitute a small current registered on the sensitive ammeter. If the 400 V supply is reversed, the current stops because electrons are not attracted to a negative electrode. Turning off the 6 V supply also reduces the current to zero.

The bulb has to be evacuated to prevent electrons colliding with air molecules. Streams of electrons moving from cathode to anode are known as *cathode rays*.

Properties of cathode rays
a) They travel in *straight* lines.
b) They are deflected by a *magnetic* field.

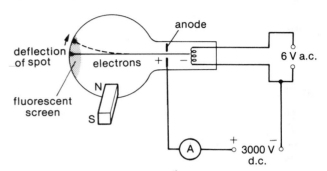

The electrons move from cathode to anode but conventional current flows in the opposite direction. In this case, the electrons are moving so quickly that they go through a hole in the anode and on to the screen.

As the magnet is brought up (with its N-S axis perpendicular to the plane of this page) the beam of electrons is deflected upwards. The deflection is seen as an upward movement of the spot on the fluorescent screen.

Fleming's left-hand rule applies here: conventional current is moving from left to right in the diagram.
c) They are deflected by an *electric* field.

The two deflecting plates have an electric field between them and as the electrons move through the field, they are deflected downwards towards the positive plate.

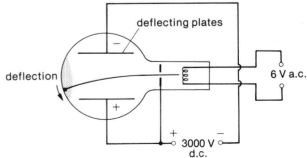

The direction of the deflection can be altered by changing the direction of the field or by altering the orientation of the plates. This principle is used in the cathode ray oscilloscope:

Cathode ray oscilloscope (CRO)

The structure of the CRO is shown in the diagram below.

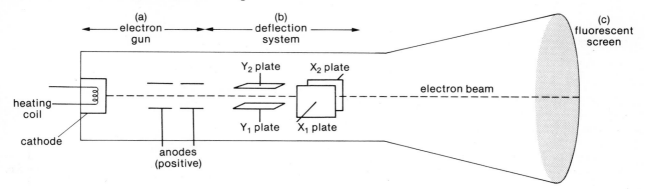

a) The electron gun

When the cathode is heated, it emits a stream of electrons by thermionic emission. The electrons are accelerated towards the anode and are focused into a narrow beam.

b) The deflection system

The Y-plates cause a vertical deflection of the electron beam. If Y_2 is positive with respect to Y_1 the beam will be deflected upwards and vice versa.

The X-plates cause a horizontal deflection of the beam in the same manner.

c) The screen

The screen is coated with zinc sulphide so that fluorescence occurs where the beam strikes it and a green spot is seen.

Uses of the CRO

When the CRO is being used, a special circuit (called a time base) is connected to the X-plates. When the time base is switched on, the spot is drawn horizontally across the screen repeatedly. To make this happen a special kind of fluctuating voltage, called a saw-tooth voltage, is applied to the X-plates. This increases slowly to a maximum value and then quickly falls to zero.

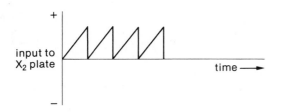

The speed with which the spot moves across the screen can be controlled. If it moves quickly a continuous line is seen on the screen.

a) Measuring a d.c. voltage

If a d.c. electric voltage is applied to the Y-plates while the time base is in operation, the following type of trace can be seen.

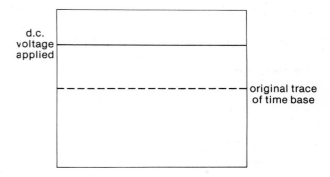

The deflection of the time base line is proportional to the applied d.c. voltage.

b) Measuring an a.c. voltage

If an a.c. voltage is applied to the Y-plates while the time base is in operation, the following type of trace can be seen.

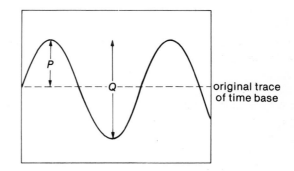

In this case the peak value P can be measured or the peak-to-peak value Q. Again deflection is proportional to applied voltage.

c) Measuring current

A CRO can be used to measure current if it is connected in a circuit across a resistor. The CRO measures the voltage across the resistor and Ohm's law is used to find the current through it.

d) Investigating waveforms

Voltage waveforms from a low voltage source can be studied using a CRO. For example, the CRO will display the trace of a sound wave when a microphone is connected to the Y-plates. A voltage variation is produced as the sound energy is converted to electrical energy and a trace is shown on the CRO screen.

e) Measuring frequency

The frequency of a particular waveform can be measured by comparing the trace of the unknown frequency with that of a known frequency. By altering the time base appropriately, the number of complete waves in a given distance can be compared.

Photo-electric emission

Electrons may be emitted from the surface of a metal when electromagnetic radiation of sufficiently low wavelength falls upon it.

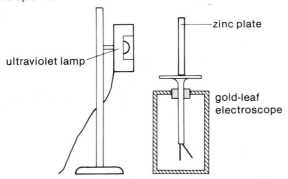

The zinc plate, which has been cleaned with emery paper, sits on top of a negatively charged electroscope. When the ultraviolet lamp is switched on, the electroscope discharges. Electrons are given off from the zinc and they are encouraged to leave by the negative charge on the electroscope. The electrons on the gold leaf of the electroscope move up to take the place of those that have left and the electroscope is discharged.

The electroscope does not discharge if it is positively charged in the first place; the positive charge prevents the departure of the electrons.

Exercises

1. What is thermionic emission? How can it be increased?
2. Where do cathode rays come from and where do they go to?
3. Give three properties of cathode rays and outline an experiment to demonstrate one of them.
4. Name the parts of the CRO labelled 1–5.

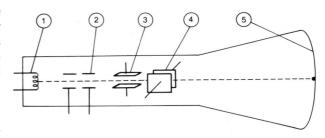

5. Explain what is meant by the 'time base' on the CRO.
6. How would you adjust the controls of the CRO to make a single spot in the middle of the screen show each of these traces?

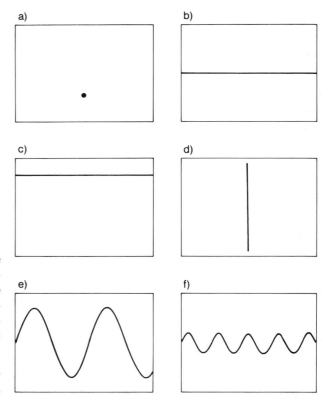

7. How would you demonstrate photoelectric emission?
8. Why do you think it is necessary to clean the zinc plate carefully before demonstrating photoelectric emission? What do you think would be the effect of putting a sheet of glass in between the zinc plate and the U.V. lamp?

7.2 The nucleus

Every atom consists of a small positively charged *nucleus* surrounded by a fuzzy cloud of negative electrons (a hydrogen atom has only one electron).

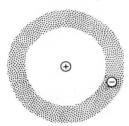

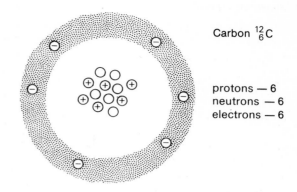

Carbon $^{12}_{6}C$

protons — 6
neutrons — 6
electrons — 6

There are two kinds of particle that are found in the nuclei of atoms, *protons* and *neutrons*.

Protons carry a positive charge equal in size to the charge on an electron but of the opposite sign. A proton has a mass equal to the mass of 1836 electrons added together.

The nucleus of a hydrogen atom is a single proton. The number of protons in the nucleus of an atom is called the *atomic number Z* of the element. So the atomic number of hydrogen is one. The atomic number of helium is two because it has two protons in the nucleus. In a neutral atom the number of protons in the nucleus is equal to the number of electrons.

Neutrons have no charge at all and have a mass roughly equal to that of a proton. There are no neutrons in the nucleus of a hydrogen atom but there are two or more neutrons in the nucleus of every other kind of atom.

The total number of protons and neutrons together in the nucleus of an atom is called the *mass number A* of the element.

Here are some examples of atoms:

The diagrams show the shorthand notation used to describe each atom,

e.g. $^{12}_{6}C$

The upper number is the mass number and the lower number is the atomic number. The symbol is the normal chemical symbol for the element concerned.

Isotopes

The chemical properties of an element are determined by the number and arrangement of its electrons. Some atoms have the same chemical properties but have different atomic masses. Such different forms of atoms are known as *isotopes*.

Isotopes of the same element have the same atomic number but a different mass number, i.e. they have the same number of protons but a different number of neutrons. For example, there are three isotopes of hydrogen:

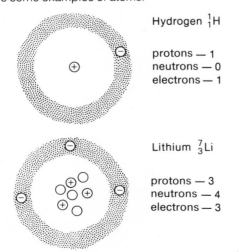

Hydrogen $^{1}_{1}H$

protons — 1
neutrons — 0
electrons — 1

Lithium $^{7}_{3}Li$

protons — 3
neutrons — 4
electrons — 3

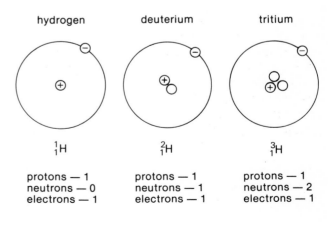

hydrogen   deuterium   tritium

$^{1}_{1}H$ $^{2}_{1}H$ $^{3}_{1}H$

protons — 1   protons — 1   protons — 1
neutrons — 0   neutrons — 1   neutrons — 2
electrons — 1   electrons — 1   electrons — 1

Radioactivity

Some isotopes are unstable and spontaneously change the structure of their nuclei, emitting radiation as they do so. This process is called *radioactivity*.

There are three kinds of radioactive emission as follows.

Alpha, beta and gamma radiation

a) Alpha (α) radiation consists of positively charged particles called α-particles. Each α-particle has the same structure as a helium nucleus: two protons plus two neutrons. α-radiation has a penetrating range of only a few centimetres in air and is easily stopped by a sheet of paper. α-particles can *ionize* air molecules by breaking the neutral atoms in the molecules into separate, charged particles (ions).
b) Beta (β) radiation consists of a stream of fast moving β-particles. Each β-particle is an electron. β-radiation also ionizes air and can penetrate several millimetres of aluminium. Its penetrating power is greater than that of α-particles.
c) Gamma (γ) radiation is an electromagnetic radiation from the short-wavelength, high-energy end of the electromagnetic spectrum (see p. 30). γ-rays travel with the velocity of light and have no charge. Consequently they are very penetrating and can pass through several centimetres of lead.

The three radiations may be distinguished from one another by their penetrating power or by their deflection in a magnetic field.

Deflection in a magnetic field

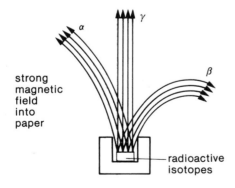

The magnetic field is directed into the plane of the paper and the radiations are deflected as shown. The movements follow Fleming's left-hand rule; γ-radiation has no charge and is not deflected.

Biological effects of radiation

Radioactive materials must be handled very carefully as the radiation that they give off is harmful to living tissue. As well as destroying cells, relatively low doses of radiation can also induce changes which eventually lead to the development of cancer. High doses of radiation are used to kill cells that are already cancerous.

Radioactive isotopes are used as 'tracers' in medicine to monitor the movement of atoms around the human body and to detect the presence of growths.

Radioactive decay

The atoms of a radioactive element decay by emitting an α- or a β-particle. As a result of this, the nucleus of the atom can become 'excited'; in settling down to a stable state, γ-rays may also be emitted.

a) α-decay is the emission of an α-particle. When this happens two changes occur:

   i) the mass number of the decaying atom goes down by 4;
   ii) the atomic number goes down by 2.

Since the atomic number is changed, the atom becomes an atom of another element (a process called *transmutation*).

For example:

$$^{226}_{88}Ra \rightarrow {}^{222}_{86}Rn + {}^{4}_{2}He$$
radium   radon   α-particle

b) β-decay is the emission of an electron from the nucleus of the decaying atom. In effect, a neutron in the nucleus becomes a proton and an electron; the electron is ejected and we call it a β-particle. The atomic number of the nucleus is increased by one and again the atom becomes one of a different element. There is no change in mass number as the mass of a β-particle is negligible compared to the mass of the whole atom involved in the decay process. For example:

$$^{14}_{6}C \rightarrow {}^{16}_{7}N + {}^{0}_{-1}e$$
carbon   nitrogen   β-particle

Half-life

Radioactive decay is a random process; it is impossible to say when an individual atom will disintegrate, although large amounts of the atoms taken together have a definite decay rate. A useful way of thinking about the decay rate of an isotope is in terms of its *half-life*. The half-life is the time taken for half the atoms in the sample to decay.

For example, 100 g of a radioactive element has a half-life of two days. After two days there will be 50 g of the radioactive element left. After four days there will be 25 g left and after six days only 12.5 g remaining, and so on.

Note that the process is continuous and does not happen in jumps. Also there is little loss of mass because the rest of the 100 g is made up of decay product.

Half-life can be found from a graph of count-rate against time. Counts are usually provided by a Geiger tube and counter; the graph produced may look like this.

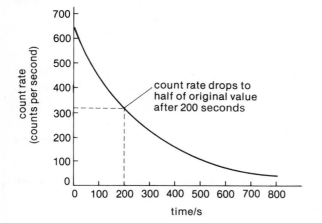

Exercises

1 Where is most of the mass concentrated in any atom?

2 Two isotopes of chlorine are $^{37}_{17}Cl$ $^{35}_{17}Cl$. Explain what the word 'isotope' means and the significance of the numbers next to the symbols for chlorine.

3 Draw up a table summarizing the facts about α-, β-, and γ-radiation under the headings: nature, charge, mass, penetration, and deflection in magnetic field.

4 If $_{-1}^{0}e$ is the symbol for an electron (or a β-particle), suggest symbols for the proton and the neutron.

5 Draw a diagram representing the isotope $^{14}_{7}N$ showing protons, neutrons and electrons.

6 Which radiation do you think is most dangerous to man, α-, β- or γ-radiation and why?

7 What effect does α-decay have on the mass number and atomic number of the decaying atom?

8 Copy out and fill in the missing numbers in those equations:

a) $^{232}_{90}Th$ → $^{?}_{?}Ra$ + $^{4}_{2}He$
   thorium   radium   α-particle

b) $^{228}_{89}Ac$ → $^{?}_{?}Th$ + $^{0}_{-1}e$
   actinium   thorium   β-particle

c) $^{224}_{88}Ra$ → $^{?}_{?}Rn$ + $^{4}_{2}He$
   radium   radon   α-particle

9 A uranium isotope decays to a thorium isotope by α-emission and has a half-life of 4.5×10^9 years. What does 'half-life' mean?

10 Radon ($^{220}_{86}Rn$) decays to polonium ($^{216}_{84}Po$) by the emission of α-particles. An experiment was set up to measure the half-life of $^{220}_{86}Rn$ and here are the results obtained.

counts per second	time elapsed in seconds
19 308	0
15 462	20
11 989	40
8 656	60
6 852	80
5 200	100
3 143	140
1 734	180
1 134	220
496	300

Draw a graph of counts per second against time elapsed and find the half-life of radon 220.

Examination questions

The publisher thanks the following Examination Boards for permission to reproduce questions from past examination papers: Associated Examining Board [A.E.B.]
Joint Matriculation Board: GCE O level [J.M.B. O]
Joint Council for 16+ (J.M.B. 16+)
Oxford Delegacy of Local Examinations [Oxford]
Southern Regional Examination Board [S.R.E.B.]
University of London Entrance and Schools Examination Council [London]
Yorkshire and Humberside Regional Examinations Board [Y.R.E.B.]

1 The diagram represents a rectangular block of iron having the dimensions indicated.

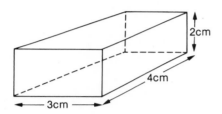

a) Calculate the volume of the block.
b) Calculate the density of iron given that the mass of the block is 0.168 kg.
c) Calculate the weight of the block.
d) If the block is placed on a bench with its face of greatest area in contact with the bench, calculate the pressure exerted on the bench by the block.
e) What is the force exerted by the bench on the block of iron?

[J.M.B. 16+]

2 A room measuring 8 m by 5 m by 3 m is full of air, of density 1.2 kg/m³.
a) What is the volume of the air in the room?
b) What is the mass of the air in the room?
c) What is the weight of the air in the room?
d) The room is open to the atmosphere in which conditions are constant. State and explain the changes, if any, that take place in the pressure and density of air in the room when the room temperature rises.

[Oxford]

3 a) The diagram shows one form of a hydrometer to be used to measure the density of liquids over the range 0.75 g/cm³ to 1.05 g/cm³.

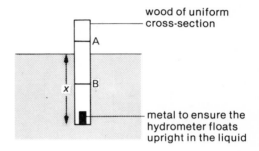

The wood is 10.0 cm long and has a cross-section area of 1.5 cm². The marks A and B indicate the limits of the scale on the hydrometer.
  i) Which of the two densities 0.75 g/cm³ and 1.05 g/cm³ would the mark B represent?
  ii) If the depth of immersion x when floating in a liquid of density 0.80 g/cm³ is 7.5 cm, calculate the mass of the hydrometer.
  iii) Calculate the distance of the mark A from the base of the hydrometer.
  iv) State the physics principle you have used in parts ii) and iii).
  v) A student using this hydrometer decided to mark a scale between the limits A and B. Would he be correct in doing this? Give a reason for your answer.
b) Water, after falling through a height of 16 m, enters a turbine at a rate of 5 kg/s. The wheel is used to lift sacks of grain through a vertical distance of 8 m.
  i) Calculate the energy given to the turbine each second if the water has negligible velocity on leaving the turbine.
  ii) Assuming that no energy is lost, calculate the maximum mass of grain which can be lifted in one minute.

[A.E.B.]

4 a) State Hooke's law concerning the behaviour of a helical spring or wire acted upon by a stretching force.
b) Describe an experiment to show that Hooke's law applies to a helical steel spring. Your answer should include
  i) a diagram of the apparatus,
  ii) an account of the observations taken,
  iii) an account of how these observations would be used to deduce the final result.
c) You are provided with a steel spring and one 1 kg mass and told to calibrate the spring as a balance reading up to 1 kg.
  i) How would you make sure that the spring is suitable for this range of loads?
  ii) Assuming that the spring is suitable, describe how you would complete the calibration.
d) A spring has a length of 10 cm when it is unloaded and 25 cm when a mass of 1.5 kg is hung on the end.
  i) Calculate the force required to extend the spring by 1 cm.
  ii) What force would be needed to make the spring 30 cm long if Hooke's law still applies?
[Assume that the acceleration of free fall (due to gravity) is 10 m/s² (N/kg).]
[J.M.B.O]

5 a) i) What is meant by the weight of a body?
  ii) Which piece of equipment would you use to measure the weight of a body?
  iii) What is meant by the mass of a body?
  iv) Which piece of equipment would you use to measure the mass of a body?
b)

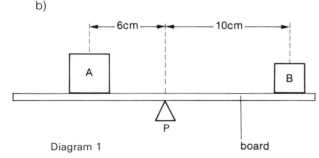

Diagram 1   board

In diagram (1) the uniform board is pivoted at its mid-point P and has two cubes A and B resting on it in the position shown. The board is in equilibrium.
  i) Explain the meaning of the word uniform in this example.
  ii) Explain the meaning of the phrase 'in equilibrium' in this example.
  iii) If the mass of the cube A is 500 g, calculate the mass of the cube B.

c) Cube B is now taken off the board and replaced by a third cube C in the same position as B. The board tilts as shown in the diagram (2) below.

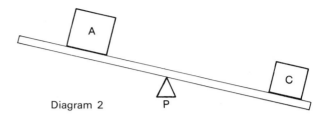

Diagram 2   P

Comparing the mass of cube C with that of cube B, this shows that the mass of cube C is _____ than the mass of cube B. Write down the missing word.
d) Copy the table below and fill in the gaps using your results from part (b), (iii).

cube	mass (g)	length of a side (cm)	volume (cm³)	density (g/cm³)
A	500	5		
B			27	
C		4		6

[S.R.E.B.]

6 The pulley system shown below has two wheels in each block and requires an effort of 12 N to lift a load of 4 kg at constant speed.

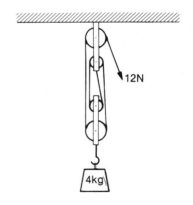

a) Define the terms *mechanical advantage* and *velocity ratio*. Find their values in this case.
b) Calculate the work to be done by the effort to raise the load 2 m and the efficiency of the machine.
c) What would be the effect, if any, on the mechanical advantage and the velocity ratio, if the machine were well oiled to reduce friction? Explain.
d) How can we tell that the weight of the movable pulley block is less than 8 N?

[Oxford]

7 a) Calculate the speed in km/h which is equivalent to 25 m/s.

b) A motor car of mass 900 kg accelerates uniformly from rest, travels at a constant speed and is then brought to rest by a constant braking force of 1800 N. The speed of the car, recorded up to the time when the brake was applied, was as follows:

time in s	0	2	4	6	8	10	12	14
speed in m/s	0	6	12	18	24	24	24	24

Using 1 cm to represent 2 seconds and also to represent 2 m/s, draw a speed-time graph to illustrate the speed of the car up to the time that it stopped.

Calculate
  i) the acceleration of the car during the first 8 seconds,
  ii) the magnitude of the accelerating force during the first 8 seconds,
  iii) the distance travelled during the first 8 seconds,
  iv) the distance travelled at constant speed,
  v) the magnitude of the momentum of the car when it was travelling at constant speed,
  vi) the total distance travelled by the car until it stopped.
[A.E.B.]

8 a) Define *momentum, kinetic energy*. State the *law of conservation of momentum*.

b) Two wheeled trolleys, X (mass 3 kg) and Y (mass 4 kg) which can run on horizontal rails, are held together at rest against a compressed spring (see diagram below). When they are released at the same instant, X moves to the left at 8 m/s.

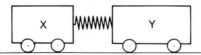

Calculate:
  i) the momentum of X immediately after release;
  ii) the momentum of Y immediately after release;
  iii) the velocity of Y immediately after release;
  iv) the kinetic energy of X and the kinetic energy of Y immediately after release;
  v) the potential energy that was stored in the compressed spring.

By the time the trolleys have come to rest again, all their kinetic energy has been converted 'doing work against friction'. Where does the friction occur and what is the final form of the converted energy?
[Oxford]

9 a) A block is released and allowed to slide down a rough slope.
  i) Describe the motion of the block by referring to the forces acting on it and the energy changes which take place during the slide.
  ii) If the slope were well-lubricated, describe what differences in the motion there would be compared with those described in (a) (i); account for the differences mentioned.

b) The slope mentioned in (a) has a vertical height of 0.5 m and measures 1.5 m along the slope. The block has a mass of 2 kg and takes 2.5 s to slide the whole distance down the slope.

Calculate
  i) the total energy lost by the block and
  ii) the average power developed.

c) The same block as in (a) and (b) is allowed to fall freely from rest from a vertical height of 5.0 m and penetrates a distance of 0.02 m into soft sand at the end of its fall.

Calculate
  i) the velocity of the block as it strikes the sand,
  ii) the kinetic energy of the block as it strikes the sand,
  iii) the average force exerted on the block by the sand in bringing the block to rest.

d) Name *two* practical devices which depend on the change of potential energy into kinetic energy.
[Assume that the acceleration of free fall $g = 10$ m/s² (N/kg).]
[J.M.B.O]

10 a) i) Explain the differences between a scalar and a vector quantity.
  ii) A tug-of-war team is composed of physics teachers who challenge two teams (team A and team B) of pupils at the same time. The teachers insist that the pupils' team pull in the directions shown on Figure 1. The teachers' rope is attached to the pupils' rope at X.

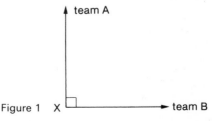

Figure 1

If team A exerts a pull of 3000 N and team B exerts a pull of 4000 N, by using a scale diagram or by calculation find the size and direction of the pull exerted by the teachers so that none of the ropes move.

b) Figure 2 gives a plan view of a small body (indicated by the dot at the top of the diagram) tied to a piece of string (indicated by the line from this dot to the centre of the circle). The body is rotated in a *horizontal* circle at a constant speed of 5 m/s. The body is 1.25 m above the ground.

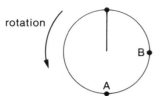

Figure 2

i) Copy Figure 2 and draw arrows to show the velocity of the body when it is at position A and then when it is at position B.
ii) Sketch a separate labelled diagram to show the change in velocity of the body as it moves from A to B.
iii) Copy and complete the following sentence. The body is moving in a circle of constant speed so that the direction of its acceleration is _____, and the force causing this acceleration is called the _____ force.
iv) The string suddenly breaks when the body next reaches position A. Assuming that the acceleration due to gravity $g = 10$ m/s^2 (i.e. gravitational field strength $= 10$ N/kg) and that frictional forces are neglected answer the following.
  A In what direction does the body move initially?
  B How long does it take for the body to reach the ground?
  C What horizontal distance does the body travel while falling?
  D Sketch the path followed by the body as it falls.

[J.M.B.16+]

11 a) The diagram below shows a wave shape. Make a copy of it.

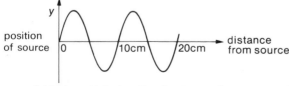

i) Mark and label on the diagram a distance equal to the wavelength of the wave.
ii) Mark and label on the diagram a distance equal to the amplitude of the wave.
iii) What physical quantity could be represented by the *y* axis if the wave is: (A) a water wave, (B) a sound wave in air?

b) The table indicates some members of the electromagnetic spectrum.

short wavelength			visible light			long wavelength
	ultra-violet rays					red light

i) What is meant by the electromagnetic spectrum?
ii) Copy the table and fill in the empty spaces in the table from the following list (which is not in order).

  blue light infrared rays green light
  radiowaves X-rays

iii) Give one example of the use of the reflection of sound waves.
Give one example of the use of the reflection of radio waves.

c) During a thunderstorm the sound of thunder is first heard about five seconds after a lightning flash. The speed of sound is 340 m/sec.
i) Explain why the thunder is heard after the lightning is seen.
ii) Estimate how far from the observer the flash occurred.
iii) Why is the speed of light not taken into account in the calculation?

[S.R.E.B.]

12 a) Diagram 1 represents a set of parallel waves travelling towards an obstacle in which there is a narrow slit opening (or 'aperture') of width *x*. Copy and complete the diagram to show what happens to the waves after they pass through the aperture.

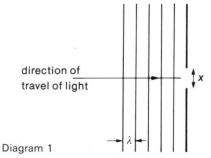

Diagram 1

What name is given to this phenomenon?
What is the effect of both increasing and decreasing the width *x* using waves of a particular wavelength λ? (You may either write statements, or draw diagrams, or both.)

b) i) What do we call the effect which results from the combination of the waves after they have passed through the double slit arrangement illustrated in Diagram 2?

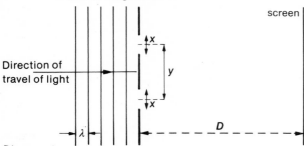

Diagram 2

ii) If the waves used in the arrangement represented by Diagram 2 are light waves of a single wavelength λ (i.e. monochromatic light), describe and *explain* with the aid of a diagram the effect produced on the screen placed at a distance *D* from the slits, as shown.
iii) For a given distance *y* between the slits, what would be the effect on what is observed on the screen if the colour of the light were changed from, say, red to green?
What does this indicate about the difference between red and green light?
iv) What do you think would be the effect (using light of one colour) of increasing *y*?
v) What do you think would be the effect of increasing *D* (again, keeping the same colour light)?

[J.M.B.16+]

13 a) What is it that all sound-producing sources such as loudspeakers and musical instruments do which causes them to emit sound energy?
b) With the aid of a diagram describe how the sound reaches a listener.

[J.M.B.O]

14 a) Diagram 1 shows a wave form as it might be displayed by a cathode ray oscilloscope.

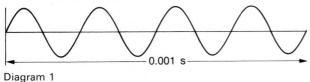

Diagram 1

i) How many complete cycles are shown?
ii) What is the frequency of the wave form shown?
iii) Draw a diagram to show a wave of twice the frequency and half the amplitude of the one shown in Diagram 1.
iv) A radio wave has a frequency of 3 MHz and travels with a velocity of 3×10^8 m/s. What is its wavelength?

b) Diagram 2 shows two wires of equal length stretched by weights. The length of the wires free to vibrate can be adjusted by moving the bridges. Wire B has twice the diameter of wire A.

Diagram 2

i) Without moving the bridges and without using additional apparatus, how could the pitch of the note produced by wire A be raised and that produced by wire B lowered?
ii) By adjustment of the tension and length, wire B is made to produce a note of 400 Hz. If wire A were then adjusted to produce a note of pitch exactly one octave higher, what would be the frequency of this note?

c) i) What name is given to a sound wave to distinguish it from the type of wave which can be seen as a ripple on the surface of water?
ii) Copy and complete the following statements which compare sound waves and electromagnetic waves:
A sound wave travels at a _____ velocity than electromagnetic waves.

Sound waves are unable to travel through _____ but electromagnetic waves can.

The wavelength of most electromagnetic waves are considerably _____ than those of sound waves.

[S.R.E.B.]

15 Light, infrared radiation, ultraviolet radiation, radio waves, X-rays and γ-radiation are all called *electromagnetic waves*.
a) State one property which all of them possess, which gives evidence that they are wave motions. Describe a simple experiment which demonstrates this property for visible light.
b) State one property which gives evidence that they are all transverse wave motions and explain why the property is evidence for this.
c) Starting with the shortest wavelength on the left, draw a diagram showing the spectrum of the radiations arranged in increasing order of wavelength.
d) Which radiation (or radiations) can be received when transmitted from a point on the opposite side of the Earth? How is this possible?
e) Which radiation (or radiations) can ionize air and other gases? Give a reason for this property.

[Oxford]

16 Two waves, A and B, interfere. Their displacement-time graphs are shown below. On the first diagram the resultant has been added.

a) Draw in the resultant on the remaining diagrams after you have copied them.

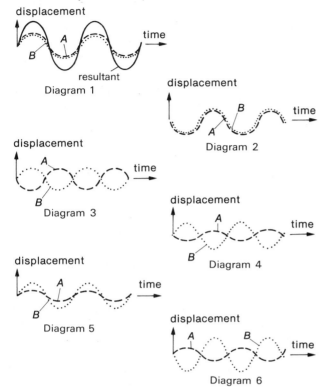

b) The waves in Diagram 1 are 'in phase', i.e. 'in step' with each other.

Which of the other diagrams represent(s) two waves in phase?

Which of the diagrams represent(s) waves arriving in antiphase, i.e. completely out of step?

Which of the diagrams represent(s) waves of different amplitudes from each other?

If the waves are light waves, which diagram represents a condition of greatest brightness?

Which diagram represents a condition of darkness?

c) Using the ideas dealt with in parts a) and b), explain in simple terms
  i) why a diffraction grating may give rise to several images of a bright, distant, monochromatic source of light, depending on the angle of view through the grating.
  ii) how a grating may produce a spectrum from white light.

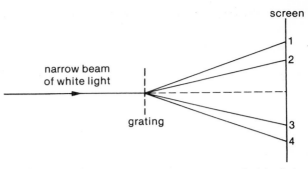

d) The diagram represents two spectra of white light produced by a diffraction grating on a screen. If the numbers 1, 2, 3 and 4 represent the limits of the visible spectra, what colours would be represented by these numbers?

e) If a grating is replaced by one having a greater number of lines per centimetre, what effect has this on
  i) the appearance of the spectrum from a white light source?
  ii) the number of such spectra which the grating may produce?

[J.M.B.16+]

17 a) Sound travels through the air as a succession of compressions and rarefactions, which constitute a longitudinal wave-motion. Explain this process in simple terms.

b) Two vertical walls A and B (see diagram) are 55 m apart. A man standing at P, 22 m from A, claps his hands once.
  i) What is the time-interval between the clap and the first echo that he hears?
  ii) What is the time-interval between the clap and the second echo?
  iii) Why does the man standing at P hear a sequence of echoes which gradually dies away?
(Take the speed of sound in air to be 330 m/s.)

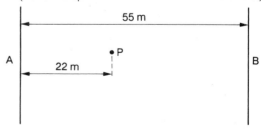

c) A stretched wire 1.2 m long vibrates with fundamental frequency 262 Hz. A bridge placed 0.4 m from one end divides the wire into two separate sections, without altering the tension appreciably. What is the fundamental frequency of each section?

[Oxford]

18 The diagram shows a stretched string which has been set into vibration by plucking. This produces a stationary wave arising from the superposition of two progressive waves that travel to and fro along the string in opposite directions and are repeatedly reflected at the fixed ends.

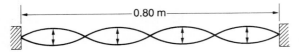

a) List the main differences between a stationary wave and a progressive (travelling) wave.
b) What is the wavelength of the waves travelling along the string? Given that the frequency is 440 Hz, what is the speed of these waves?
c) Draw a diagram showing the pattern of vibration for a higher frequency at which the string can vibrate, and state the value of this frequency.
d) Draw a diagram showing the vibration pattern for the lowest frequency at which the string can vibrate, i.e. the fundamental, and state the value of this frequency.
e) If you were provided with a variable frequency sound source which is calibrated to emit pure tones over a wide range of frequencies, how could you measure the fundamental frequency of such a string directly?

[Oxford]

19 A ship's echo sounder sends short pulses of sound waves down from a sender unit, A, and they are collected by a receiver unit, B. The frequency used is 20 000 Hz. The speed of sound in water is 1500 m/s. It takes 0.1 s for the sound from A to reach B.

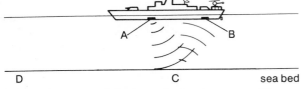

a) What is happening to the sound waves at C?
b) Calculate the depth of water beneath the ship.
c) Calculate the wavelength of these sound waves in water.
d) As the sound waves travel from A to C, what changes are there, if any, in
   i) their frequency?   ii) their amplitude?
e) Explain why it would be difficult to use this method to find the depth of water in front of the ship at D.
f) In theory, it should be possible for aircraft to use sound echoes to measure their height above the ground. Give **one** reason why this is not done.

[Y.R.E.B.]

20

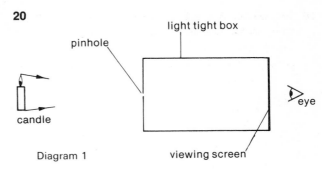

Diagram 1

a) Diagram 1 represents a simple pinhole camera used to look at a lighted candle.
   i) Complete the diagram showing how the rays from the top and the bottom of the candle pass through the camera.
   ii) Show how the candle's image would appear on the viewing screen.
b) What information does this experiment show concerning the way in which light travels?
c) i) If the pinhole is made larger, what two effects will this have on the image?
   ii) If the distance between the pinhole and the candle is doubled, what two effects will this have on the image?
d) i) If three pinholes are made instead of one, then three images are produced. How would you use a lens to bring the images together in one place on the screen?
   ii) Would the lens be convex or concave?
e) Three parallel rays of light are incident separately on:
   a parallel sided glass block
   a convex (converging) lens of short focal length
   a convex lens of long focal length and
   a concave (diverging) lens
   State which piece of optical equipment should be inserted in the gap in each of the diagrams below.

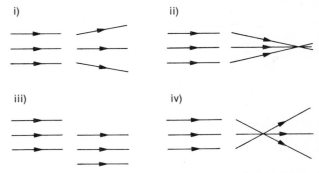

Copy and complete the diagram of the rays incident on the parallel-sided glass block showing how these rays pass through the block.

[S.R.E.B.]

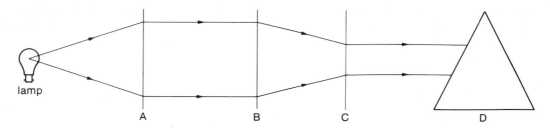

21 The diagram above represents two rays of light from a lamp passing through a series of lenses followed by a glass prism. The diagram is drawn full size. Make a copy of it.
  a) At line A draw the shape of the lens which must be situated on line A.
  b) Name the type of lens used at line B.
  c) Name the type of lens used at C.
  d) Measure and write down the focal length of the lens at A.
  e) Measure and write down the focal length of the lens at B.
  f) Copy the diagram, and draw and label a set of rays to show the effects of refraction and dispersion on the two rays shown as they pass through the prism.
  g) On the copy of the diagram, draw a dotted line and label it M to show where a plane mirror could be placed in order to reflect both rays back to their starting point.

[Y.R.E.B.]

22 a) List *four* properties of the image of a real object formed by a plane mirror.
  b) A plane mirror is often used in an ammeter to assist accuracy in reading the scale.
  i) Draw a diagram to show how a mirror is arranged in relation to the scale and the pointer.
  ii) How is the mirror used to obtain an accurate reading on the scale?

[J.M.B.O]

23 a) Draw ray diagrams to show how i) a convex (converging) lens and ii) a concave (converging) mirror can produce a *virtual image*.
  Suggest a practical use for each of the above two cases.
  b) Two defects of the eye are hypermetropia (long sight) and myopia (short sight).

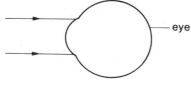

Copy this diagram.

Choose one of the two defects. State which one you have chosen and show what happens to the two rays of light when they enter the eye which has the defect you have chosen. State how you would correct this defect.
  c) The diagram below shows the arrangement (drawn to scale) in an optical instrument.

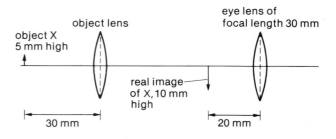

Find by calculation or by construction on graph paper:
  i) the focal length of the object lens,
  ii) the position and size of the final image produced by the eye lens.
(If you solve the question by calculation, state the sign convention used.)

[A.E.B.]

24 a) i) Draw a ray diagram showing how a lens may be used as a magnifying glass.
  ii) Name the type of lens used.
  iii) Describe the image formed by the lens used in this way.
  iv) Suggest a suitable value for the focal length of the lens you might usefully use in this way.
  v) Calculate the power of the lens referred to in part (iv).
  b) A lens used as in a) is often used, together with another lens, to form an astronomical telescope.
  i) Draw a ray diagram of such a telescope, name the types of lens used and show the position of the principal foci of the lenses and the positions of
  the images formed.
  ii) Describe the images formed by the lenses used in this way.

iii) Suggest a suitable value for the focal length of the objective lens you would use.

[J.M.B.O]

25 a) Describe how you would determine accurately the focal length of a converging lens using a laboratory method. (Assume that you cannot see the Sun or any distant object.)

b) The focal length of a lens is found to be 10.0 cm. Find how far you would place it from an illuminated slide to obtain an image on a screen which was magnified five times.

Name two characteristics of the image (other than that it is magnified).

Suppose you drop the lens and it breaks so that only half of it is intact. You replace it in the same position to throw an image on the screen. State what effect, if any, this would have on the size and brightness of the image. (Give reasons for your answers.)

[London]

26 The diagram below shows some parts of a slide projector facing a screen. Make a copy of it.

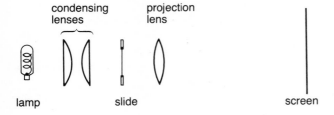

a) On the diagram, draw a suitable mirror correctly placed to increase the amount of light passing through the condensing lenses.
b) From a point on the slide, draw two rays passing through the projection lens to focus on the screen.
c) On the diagram, draw and label a suitably placed heat filter.
d) Why is a heat filter necessary?
e) Why should the slide be placed upside down in the projector?
f) What is the purpose of the condensing lenses?
g) If it is required to project a larger image on to the screen and still keep it in focus, what should be done to the positions of
   i) the screen?
   ii) the projection lens?

[Y.R.E.B.]

27 a) Diagram 1 shows a corner of a room in a bungalow kept warm by a central heating system.

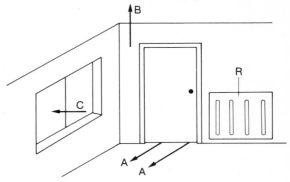

Diagram 1

i) Where the door does not quite fit its frame at the bottom, draughts of cold air (A) are found. What causes these?
ii) Heat loss through the roof can be reduced by laying certain materials between the rafters. Suggest a material which could be used and explain the particular property such materials have.
iii) Heat loss through the window (C) may be reduced by double glazing. Explain briefly how double glazing does this.
iv) R is a 'radiator' painted glossy white. Explain why, in fact, little heat is actually radiated from this and how the heat is more probably distributed through the room.

Diagram 2

b) Diagram 2 shows a method used by campers to keep milk cool. The bottle of milk is placed in a wet sock which is then hung where a breeze can reach it.
Explain carefully how this enables the temperature of the milk to be kept below air temperature.

c) 0.5 kg of a liquid at 10°C was placed in a vacuum flask and then heated for 5 minutes with a 100 W immersion heater. The temperature of the liquid rose to 25°C.
   i) How much energy was transferred in one second?
   ii) How much energy was transferred in five minutes?
   iii) What is meant by 'specific heat capacity'?
   iv) From the data given, calculate the specific heat capacity of the liquid.

[S.R.E.B.]

28 a) Choose from the following list of terms the one term in each case which fits the process described in i) to iv).

List of terms: conduction,
convection,
radiation,
insulation,
expansion.

  i) The increase in the average distance apart of molecules of a solid when heat is supplied to the solid.
  ii) The movement of hotter particles of a heated fluid away from the source of heat.
  iii) The reducing of the rate that heat passes through the roof of a house by putting thick glass fibre on the roof floor.
  iv) The heating of the atmosphere by electromagnetic waves from the Sun.

b) A bimetallic strip can be used in a thermometer (see diagram 1).

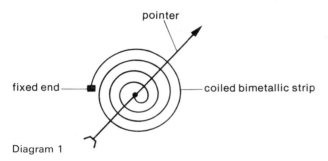
Diagram 1

  i) What is a bimetallic strip?
  ii) What materials are often used in the strip?
  iii) What would happen to the pointer attached to one end of the strip if the temperature read by the thermometer increased?

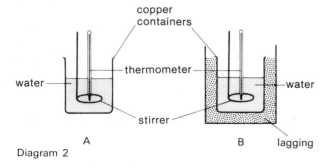
Diagram 2

c) Diagram 2 shows two identical copper containers A and B each containing the same volume of heated water, with a thermometer and stirrer. A is unlagged but B is lagged. Temperature readings are taken of the water in A and B each minute and the following readings are obtained:

time (minutes)	temperature of A (°C)	temperature of B (°C)
0	95	95
2	86	89
4	80	86
6	75	83
8	72	81

d) i) Plot on graph paper two graphs showing these observations. Use the same axes for both sets of readings and label your graph.
  ii) In which container was the rate of fall of temperature of the water greater?
  iii) From which container was the rate of loss of heat energy greater?
  iv) Why was the rate of loss of heat less in the other container?
  v) What would be a suitable material for the lagging?

[S.R.E.B.]

29 a) Heat may be transferred between bodies by conduction, convection and radiation. Describe how the energy is transferred in each case.
b) For one of the processes named in a), describe an experiment which demonstrates that energy is being transferred.
c) Explain the part played by each process in heating a room with hot water radiators.

[J.M.B.O]

30 a) Explain what is meant by (i) lower fixed point and (ii) upper fixed point. Explain their importance in setting up a temperature scale.
b) Describe how you would find the position of one of these points on an uncalibrated thermometer.
c) Sketch a graph showing how the volume of a fixed mass of water varies as it is cooled from 10°C to −5°C. Explain the significance of this graph to the freezing of freshwater ponds and streams.

[J.M.B.O]

31

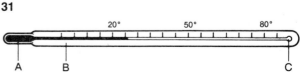

Celsius thermometer

a) Name the substance A.
b) Name the substance B.
c) Why is part C of the tube enlarged?

d) Copy the diagram and mark the freezing point of water on the thermometer with a letter F.
e) Draw and label a diagram showing how you would check the accuracy of the point F on the thermometer.
f) It takes 60 000 J of heat to raise the temperature of 500 g of porridge from 15°C to 45°C. Calculate the specific heat capacity of porridge.
[Y.R.E.B.]

32 Kinetic theory suggests that the molecules of a gas or a liquid are in continual random motion.
a) Explain in terms of kinetic theory why a gas exerts a steady pressure, which increases as the temperature rises, and which increases when the gas is compressed into a smaller space.
b) A small bottle containing liquid bromine is broken inside a larger glass vessel which has been evacuated. At once a brown coloration is seen throughout the larger vessel and the volume of the liquid is considerably decreased. After a short while the intensity of the coloration and the quantity of liquid remain constant. Explain these observations in terms of the movements of bromine molecules.
c) Explain in terms of kinetic theory why evaporation produces cooling.
d) Give a brief account of the manner in which heat is transferred by i) conduction, and ii) convection.
[Oxford]

33 a) Explain what is meant by the specific heat capacity of a substance.
b) In an experiment to measure the specific heat capacity of paraffin oil, 110 g of paraffin oil was contained in a copper calorimeter whose mass was 50 g. A 24 watt electric heater caused the temperature to rise from 10°C to 40°C in 5 minutes. Assuming that the calorimeter was well lagged, calculate
  i) the amount of heat supplied by the heater during the experiment.
  ii) the amount of heat absorbed by the calorimeter during the experiment,
  iii) the specific heat capacity of paraffin oil.
  (Assume that the specific heat capacity of copper = 0.4 J/g K.)
c) Explain why a drink is cooled more by ice than by the same mass of water at 0°C.
[J.M.B.O]

34 a) The specific heat capacity of copper is 390 J/kg deg.C?
  i) How many joules of energy are needed to raise the temperature of 1 kg of copper by 1°C?
  ii) How many joules of energy are needed to raise the temperature of 5 kg of copper by 10°C?
b) In an experiment to measure the specific heat capacity of copper, a copper cylinder is heated by a small immersion heater placed in a hole bored in the cylinder. There is another hole bored in the cylinder to hold a thermometer. A stop clock is available.
  i) What four measurements need to be made in order to obtain a value for the specific heat capacity of copper, apart from a knowledge of the power supplied by the heater?
  ii) Give one reason why the specific heat capacity of copper obtained by this method is likely to be inaccurate.
  iii) Give one precaution you would take to make the value more accurate.
  iv) Give one precaution you would take to ensure safety in carrying out this experiment.
c) The graph shows the cooling of some naphthalene which was placed in a test-tube with a thermometer. The temperature of the naphthalene when timing started was 90°C.

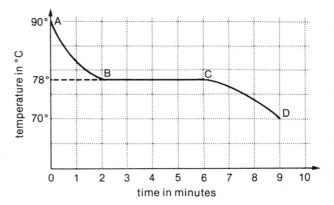

  i) What is happening to the temperature of the naphthalene from A to B?
  For parts ii) and iii) copy and complete the sentences.
  ii) The temperature of 78°C is the _____ of liquid naphthalene.
  iii) From B to C the naphthalene is changing from _____ to _____.
  iv) What is the temperature of the naphthalene at point D?
  v) What is the time taken, in minutes, from B to C?
  vi) The specific latent heat of fusion of naphthalene is 150 000 J/kg. Calculate the energy lost by 0.1 kg of naphthalene when it solidifies.
  vii) Where does this energy go to?
[S.R.E.B.]

35 a) A small glass beaker containing a thermometer is partly filled with a volatile liquid (e.g. ether) and air is bubbled violently through the liquid for several minutes. Describe what you would observe happening and explain your observations.
b) Explain what is meant by the term 'Brownian motion'.

[J.M.B.O]

36 The results shown in the table below were obtained in an experiment to verify Boyle's law.

pressure	(kN/m²)	400	320	160	80
volume	(mm³)	2.0	2.5	5.0	10.0
$\frac{1}{\text{volume}}$	(mm⁻³)	0.5			

a) Copy the table and complete it.
b) Plot a graph of pressure on the y-axis against 1/volume on the x-axis.
c) State the relationship which this graph shows between pressure and volume.
d) From your graph calculate the volume when the pressure was 240 kN/m².
e) State which two physical properties of the gas were kept constant.

[J.M.B.O]

37 a) Explain what is meant by the absolute zero of temperature. Describe a simple experiment which you could perform in your school laboratory which would allow you to estimate the value of absolute zero on the Celsius scale of temperature. Sketch the apparatus which you would use. List the observations you would make and show how these observations would be used to arrive at the final result. What result would you expect?
b) A motor car tyre contains a fixed mass of air. The pressure of the air was measured as 200 kN/m² above atmospheric pressure when the air temperature was 17°C. After a high speed run, the air pressure in the tyre was measured again and was found to be 230 kN/m² above atmospheric pressure. What was the new temperature of the air in the tyre if its volume remained constant?
(Atmospheric pressure on the day was 100 kN/m².)

[J.M.B.O]

38 a) Here is a list of materials:
nickel steel copper soft iron chromium
i) Which of these materials can be magnetized?
ii) Which of these would you use for a permanent magnet?
iii) Which of these would you use for a transformer core?
iv) 'An ordinary bar magnet has two poles, called north-seeking (N) and south-seeking (S)'. What is meant by the poles of a magnet?
v) Why are the poles called north-seeking and south-seeking?
b) i) Copy and complete the sketch, to show the pattern of iron filings formed due to two bar magnets placed in the positions shown.

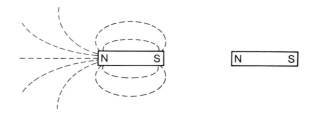

ii) The sketch below shows the pattern of iron filings formed due to two similar bar magnets in the positions shown.

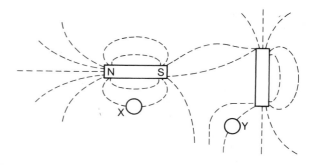

The N and S poles have been marked on only one magnet. Copy the sketch and mark the N and S poles on the other magnet. Also mark in the circles on the 'lines of force', the direction in which a compass needle would point if placed in turn at X and Y. (Ignore the effect of the Earth's magnetic field.)
c) The north-seeking pole of a strong bar magnet was brought in turn near to two magnetic materials A and B. Material A was attracted towards the magnet, but material B was repelled away from the magnet.
i) Which of A and B was definitely a magnet?
ii) Give a reason for your answer to i).
iii) What further test would have to be carried out on the other material to decide whether it is or is not a magnet?

iv) What observation would you make if this material is a magnet?
v) What observation would you make if this material is not a magnet?

[S.R.E.B.]

39 a) What are the three principal effects of an electric current?
b) Diagram 1 shows three resistors, two of 100 ohm and one of 50 ohm. Draw diagrams to show how all three resistors can be connected in different ways to give total resistances of i) 250 ohm, ii) 100 ohm, and iii) 25 ohm. Each diagram must show all three resistors.

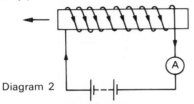

Diagram 1

c) Diagram 2 shows a coil of thick insulated wire wound around a rod made from soft iron. The battery provides 6 V and the ammeter reads 6 A.

Diagram 2

i) Copy the diagram and show beside the arrow the polarity of the magnetic field of the iron rod. The direction of the current is marked on the diagram.
ii) What is the resistance of the wire coil?
iii) How does the steel differ from soft iron, magnetically?
iv) What would be the effect on the current reading if the number of turns of wire in the coil were to be doubled?
v) What would be the effect on the magnetic field if the voltage was doubled and the number of turns of wire in the coil was also doubled?
vi) Name three common devices, other than a transformer, which depend upon electromagnetism.

[S.R.E.B.]

40 a) You are provided with a d.c. source, a switch, a variable resistor and a coil of wire wrapped around a cardboard cylinder.
i) Draw a diagram to show how you would use all the above components to magnetize a steel rod which fits inside the cardboard cylinder. On your diagram indicate the polarity of the d.c. source, the direction of the conventional current flow by ⟶ (or electron flow by ⟵e) and the magnetic polarity you would expect to obtain at each end of the steel rod.
ii) State how you have decided which end of the rod has a magnetic north pole.
iii) State one way in which the strength of the magnetism of the rod could be increased.
b) The coil of the wire is now removed from the circuit set up in a) and connected to a sensitive centre-zero galvanometer. State, giving reasons, what you would observe when the magnetized rod is
i) pushed into the coil,
ii) held stationary in the coil.
c) The coil of wire is now replaced in the circuit set up in a) and a steady current flows. The north pole of the magnetized steel rod is pushed into the north pole end of the coil. State how the current changes as the rod is pushed into the coil. Explain, in detail, why the current changes.

[A.E.B.]

41 a) A leaf electroscope is charged positively. With the aid of diagrams describe and explain what happens when
i) a positively charged rod is brought near the cap,
ii) a flat circular metal plate with an insulating handle is held close to and parallel to the cap.
iii) the bare ends of an otherwise well insulated piece of wire are connected one end to the cap and the other end to the flat circular plate.
b) Many small pieces of soft iron, identical in size and shape are held just below end A of the arrangement shown below.

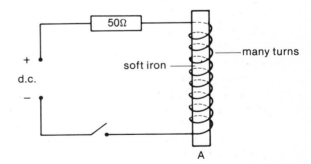

Describe and account for what happens when the switch is closed and then opened again. Compare these results with what would happen if pieces of hard steel of identical size and shape were used instead of the soft iron.

Suggest a value for the current in a practical electromagnet which is used to lift heavy sheets of mild steel.

[London]

105

42 a) With the aid of diagrams, explain what happens in terms of electron flow and the potential and divergence of the leaf of a leaf electroscope when
   i) a negatively charged rod is brought near to the top of an uncharged leaf electroscope,
   ii) the cap is then earthed by touching it with the finger,
   iii) the finger is removed from the cap and then the charged rod is moved right away,
   iv) the negatively charged rod is once again brought near the cap from a large distance.
b) The diagram below shows a charged insulated metal plate A connected to the cap of a leaf electroscope.

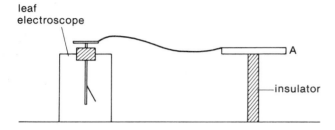

Explain the following two statements in terms of charge, potential and capacitance.
   i) If an earthed metal plate is placed directly above, but not touching, A, the deflection of the leaf in the electroscope decreases.
   ii) If the earthed metal plate is now brought closer to, but still not touching, A, the leaf deflection decreases even further.
[A.E.B.]

43 a) Draw circuit diagrams showing a set of three resistors connected as described below.
   i) all in series with each other and with a battery;
   ii) all in parallel with each other, and this combination connected in series with a battery;
   iii) two connected in parallel with each other and this combination connected in series with the third resistor and a battery
b) Calculate the total resistance across the battery in both case a) i) and case a) ii) if the resistors have values 2 Ω, 4 Ω and 6 Ω.
c) Calculate the total resistance across the battery in case a) iii) if the 2 Ω resistor is the one which is placed in series.
d) If the p.d. across the terminals of the battery in cases a) i) and a) ii) is 6.0 V, calculate
   1) the current in the circuit in case a) i),
   2) the currents in each of the three resistors in case a) ii).

e) If a current of 0.5 A flows in the 2.0 Ω resistor in case a) iii), calculate
   1) the p.d. across the terminals of the battery in case a) iii),
   2) the p.d. across the pair of resistors in parallel in case a) iii). [J.M.B.16+]

44 Two resistors, one of resistance 4 ohm and the other of unknown resistance, are connected in parallel. This combination is then placed in a circuit and the current passing into the combination is measured for various potential differences across the combination. The results of the experiment are given below.

potential difference/V	1.5	3.0	4.5	6.0	7.5
current/A	0.75	1.50	2.25	3.00	3.75

a) Draw a labelled diagram of the circuit you would use to perform the experiment. (Do not describe the experiment.)
b) i) Plot a graph of potential difference against current.
   ii) From the graph calculate the total resistance of the combination of resistors, explaining clearly how the graph was used.
   iii) Using the resistance of the combination obtained in ii), calculate a value for the unknown resistance.
c) Describe, with the aid of one diagram in each case, how a moving-coil galvanometer may be converted into i) a voltmeter, ii) an ammeter.
[J.M.B.O]

45 You are provided with the following apparatus:

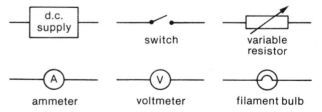

a) Draw a circuit diagram showing how you would connect the above components in order to measure the resistance of the filament for different values of the current through the filament.
b) The table below shows the ammeter and voltmeter readings for different currents through the filament.

voltmeter reading in V	0.30	0.76	1.10	2.10	3.30	4.40
ammeter reading in A	0.05	0.08	0.10	0.15	0.20	0.25

Calculate the resistance of the filament for each set of readings.

c) Plot a graph having resistance as the vertical axis and current as the horizontal axis. Use 2 cm to represent 2 Ω on the vertical axis and 2 cm to represent 0.05 A on the horizontal axis. Draw a smooth curve through your points.
d) Use your graph to calculate the resistance of the filament when the current through it is 0.18 A.
e) Use your values for current and resistance in part d) to calculate the energy dissipated by the filament in 5 minutes.

[A.E.B.]

46 a)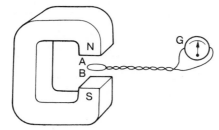

Figure 1

Figure 1 above represents a flat circular coil AB placed between the poles of a strong magnet and connected to a sensitive galvanometer, G. Explain carefully why
 i) in the position shown, with the side A of the coil uppermost, there will be no deflection of the galvanometer,
 ii) when the coil is rotated quickly through 180° so that side B of the coil is uppermost there will be a deflection,
 iii) when the coil is pulled quickly away from the magnet there will be a deflection.

b)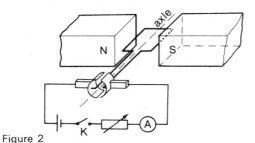

Figure 2

Figure 2 represents a coil, the ends of which are connected to a split ring commutator.

The coil is placed between the poles of a strong magnet and connected by brushes to the circuit shown. Explain why, when the key K is closed,
 i) the coil starts to rotate, and give the direction of rotation,
 ii) the coil continues to rotate.

c)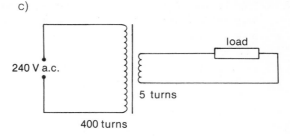

Figure 3

Figure 3 represents a transformer with 400 turns in the primary winding and 5 turns in the secondary winding. The primary e.m.f. is 240 V and the primary current is 2 A.
 i) What is the secondary voltage?
 ii) Assuming no power loss, what is the secondary current?
 iii) Energy is wasted in the transformer. Give two reasons why your calculated value in ii) will be larger than the true value of the current.
 iv) Suggest a purpose for which this transformer may be used.

[London]

47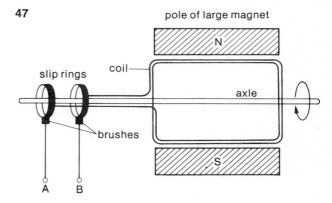

The apparatus shown in the above diagram will produce alternating e.m.f. at the terminals A and B when the coil is spun on its axis.
a) Why are slip rings and brushes used to take a current from the coil?
b) State two different ways of increasing the e.m.f. produced by the above apparatus.
c) If the coil rotates 10 times every second, what is the frequency of the a.c. which is generated?
d) State one use of electricity where it is essential to use d.c.
e) Name a device which changes a.c. to d.c.

f)

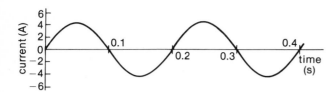

The above graph shows how the alternating current (a.c.) in a circuit varies with time.
How does the flow of electricity between 0 and 0.1 s compare with that between 0.1 s and 0.2 s?

g) i) What is the time taken for 1 cycle of this a.c.?
   ii) What is the frequency of this a.c.?

[Y.R.E.B.]

48 a) The diagram below shows a simple demonstration transformer, intended to convert a 24 V, 50 Hz a.c. supply to 240 V, 50 Hz.

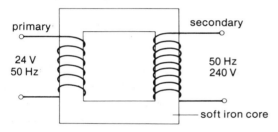

i) Why cannot such a transformer be used for steady direct current?
ii) What is the purpose of the 'soft iron' core?
iii) Why is the core usually made of sheets of a magnetic material (which has a high resistivity) that are insulated from one another?
iv) If the primary has 50 turns, how many turns should there be on the secondary? Explain briefly.
v) If the primary current is 25 A, what is the greatest possible value (in theory) of the secondary current and why will the value of the secondary current be less than this?

b) A consumer receives a power of 30 kW at 600 V at his end of the transmission lines. If the resistance of the transmission lines is 0.2 Ω calculate:
i) the current flowing in the lines;
ii) the voltage drop in the lines;
iii) the power wasted in the lines.
What would be the corresponding values for the current flowing, the voltage drop and the power wasted if the 30 kW of power were received at 60 kV instead of 600 V?

[Oxford]

49

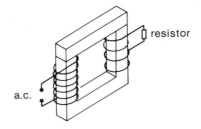

The diagram above represents a transformer with a primary coil of 400 turns and a secondary coil of 200 turns.
a) If the primary coil is connected to the 240 V a.c. mains, what will be the secondary voltage?
b) Explain carefully how the transformer works.
c) Calculate the efficiency of the transformer if the primary current is 3 A and the secondary current 5 A.
d) Give reasons why you would expect this efficiency to be less than 100%.
e)

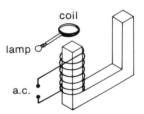

The secondary coil is removed and a small coil connected to a low voltage lamp is placed as shown. Explain the following observations:
i) the lamp lights,
ii) if the coil is moved upwards, the lamp gets dimmer,
iii) if a soft iron rod is now placed through the coil, the lamp brightens again,
iv) the lamp will not light if a d.c. supply is used instead of an a.c. one.

[London]

50 Diagrams 1 and 2 show two ways in which the current through a filament lamp may be varied when a supply voltage of V volts is available.

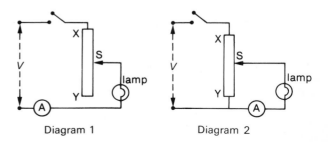

Diagram 1   Diagram 2

a) Each circuit contains a switch and an ammeter in addition to the lamp. What name is given to the other device included in both circuits?
What name do we give to the way in which this device is used in circuit 2?
Assuming that the maximum obtainable ammeter reading in either circuit represents a current through the lamp large enough to cause it to glow brightly, where should the slider S be placed to give maximum brightness – at end X or at end Y in each diagram?

b) Assuming that for all positions of S in the second circuit (diagram 2), the resistance of the lamp is relatively large compared with that of XY, describe and explain what you would expect to observe, regarding both the brightness of the lamp and the ammeter reading, as S is moved slowly from X to Y. In the first circuit (diagram 1), describe and explain what you might expect to observe regarding the brightness of the lamp and the ammeter reading as S moves slowly from X to Y, this time assuming that the resistance of the lamp is similar in value to the resistance of XY, whatever the position of S. Explain why the resistance of the lamp depends on the position of S for both circuits.
In the case of circuit 2, for what position of S would the lamp have the least resistance?

c) If a voltmeter were available, explain where you would place it in circuit 2 to perform an experiment to enable you to plot a graph of voltage against current for the lamp. State what adjustments you would make in this experiment and what readings you would need to take in order to plot the graph. Sketch the appearance of such a graph.

[J.M.B.16+]

51 a) State two practical uses of a capacitor.

b) Figure 1 shows a circuit containing a capacitor C, a sensitive ammeter A, a resistor R and a switch S. It is found that when S is closed the meter shows a decreasing reading which eventually becomes zero.

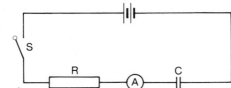
Figure 1

i) Explain why the meter reading decreases to zero.
ii) What would be the effect of replacing R with a resistor of a smaller value?

c) The capacitor C is now placed directly across an a.c. supply as shown in Figure 2.

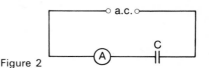
Figure 2

i) What happens to the capacitor plates during a positive half cycle of alternating voltage?
ii) What difference (if any) would occur if you had considered a negative half cycle instead of a positive half cycle?
iii) Name a suitable type of ammeter for A.

d) Figure 3 shows an inductor connected to a suitable ammeter and to an alternating current supply of a constant voltage but a varying frequency.

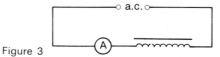
Figure 3

i) By selecting suitable words from the following list, copy and complete the paragraph below which explains the action of an inductor in an a.c. circuit.

field   flow   windings   change
aluminium   iron   copper   voltage

The inductor consists of many turns of thin _____ wire wound around an _____ core. The current through the inductor creates an alternating _____. This cuts the _____ of the inductor producing a back e.m.f. This back e.m.f. is always opposing the _____ in the current.

ii) What would happen to the ammeter reading if
   A) the core were removed,
   B) the frequency of the supply were to be lowered?

e) Figure 4 shows a circuit containing a 2 V d.c. source in series with a 6 V a.c. source together with three resistances R_1, R_2 and R_3

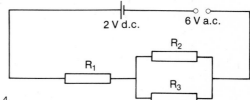
Figure 4

i) Sketch a graph of current versus time to show the current in R_1 for a few cycles of a.c.
ii) What component would you add to the circuit so that only a.c. passes through R_2?

Copy Figure 4 and using the letter X, mark the position in which you would place this component.

[J.M.B.16+]

52

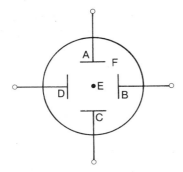

The above diagram represents a cross-section of a cathode ray tube seen from the front. E is the electron beam coming towards you. A, B, C and D are metal plates inside the tube which are used for electrostatic deflection of the beam.
a) Why is it called a 'cathode ray tube'?
b) What kind of electric charge do electrons have?
c) If plate D is connected to the positive terminal of a battery and plate B is connected to the negative terminal of the same battery,
  i) what will happen to the electron beam E?
  ii) Explain your answer to c) i).
d) Which two plates are the Y-plates?
e) Describe how one or more batteries may be connected to the plates in order to deflect the electron beam E in the direction of F.
f) Cathode ray tubes are sometimes built to use magnetic deflection instead of electrostatic deflection.
State what would be included instead of the metal plates A, B, C and D.
g) When the electrons reach the screen, most of their energy is converted to light. What causes this change?

[Y.R.E.B.]

53 a) The diagram below illustrates the structure of a cathode ray tube using electrostatic deflection.

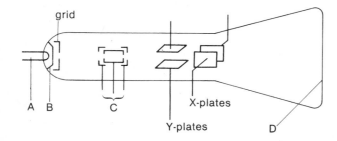

Name, and describe the function, of the components labelled A, B, C and D.
b) Using a cathode ray oscilloscope, the following results were obtained with a d.c. source connected to the Y-plates and with the time base switched off.

d.c. voltage applied to Y-plates in V	0	1.0	2.0	3.0	4.0
deflection of spot from central position in mm	0	12.0	24.5	37.0	49.5

i) Plot a graph having deflection as the vertical axis and d.c. voltage as the horizontal axis. Use a scale of 2 cm = 5 mm on the vertical axis and a scale of 2 cm = 0.5 V on the horizontal axis. Draw a straight line through your points.
ii) A d.c. source of unknown voltage was applied across the Y-plates and a deflection of 45 mm was observed. Use your graph to calculate the unknown voltage.
iii) A 50 Hz a.c. source of unknown voltage was then applied across the Y-plates and a line of length 40 mm was observed. Use your graph to calculate the peak voltage of the source, and hence calculate the r.m.s. voltage of the source. If the time base were then switched on and set at a frequency of 100 Hz, sketch what you would expect to see on the screen of the oscilloscope.

[A.E.B.]

54

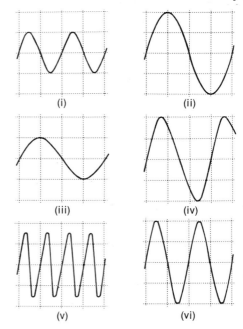

The a.c. represented by the cathode ray oscilloscope screen display illustrated in diagram (i) above, has a frequency of 60 Hz and a peak voltage of 2 V. The squares of the grid measure 1 cm horizontally and vertically.

a) Copy and complete the table below to give the values of the frequencies and peak voltage values of the other displays (diagrams (ii) to (vi)) assuming no change of settings of the controls of the oscilloscope.

diagram	(ii)	(iii)	(iv)	(v)	(vi)
frequency					
peak voltage					

b) What is the approximate r.m.s. value of the voltage shown in diagram (i)?
c) Draw a labelled circuit diagram of a circuit suitable
   i) for half wave rectification of an a.c. supply,
   ii) for full wave rectification of an a.c. supply,
   Include, in each case, a suitable smoothing arrangement connected to the output of the rectifier.

   Give a simple explanation of how the circuits function both with regard to the rectification and to the smoothing.
d) Draw diagrams to the same time base representing the shape of the trace which you would expect to see if the output from the rectifier were connected to the Y-plates of a cathode ray oscilloscope in each of the following cases:
   i) unsmoothed output from a half wave rectifier,
   ii) smoothed output from a full wave rectifier.
   [J.M.B.16+]

55 a) A nucleus can be written in symbol form as $^{218}_{84}X$.
   i) Explain the significance of the numbers 218 and 84.
   ii) This nucleus is thought to be radioactive, emitting both alpha (α) particles and gamma (γ) rays. Describe how you would test for this in the laboratory.
   iii) What would be the effect on this nucleus after emitting both of these?
b) A radioactive element is produced artificially, and its activity is measured after 2 days, 4 days and 6 days, as shown in the table below.

time in days	2	4	6
activity in counts/minute	11 000	2750	687.5

Use this information to calculate
   i) its half-life,
   ii) its initial activity.
   [A.E.B.]

56 a) Three types of nuclear radiation are alpha, beta and gamma radiation. State the type(s) of radiation (if any) which
   i) carries or carry an electric charge,
   ii) is or are *not* affected by magnetic fields.
b) For an atom define:
   i) its atomic number,
   ii) its mass number.
c) i) In what way are the isotopes of an element similar?
   ii) In what way are the isotopes of an element different?
d) i) The nuclear radiation from a source is investigated by a student using a counter which can detect alpha, beta and gamma radiation. The student first notes the count rate when the source is not present. What is the counter detecting and why does the student do this?
   ii) When the counter is placed close to the source, a high count rate is observed. When a piece of paper is placed between the source and the counter there is very little change in the count rate.

   Explain what deduction can be made about the nuclear radiation emitted by the source.
   iii) It is found that the count rate is reduced to half its original value when a thin sheet of low-density metal is placed between the source and the counter.

   Name the nuclear radiation being heavily absorbed by thin metal sheet and explain another way of checking that the source emits that radiation.
   iv) When more thin sheets of the low-density metal are placed between the source and the counter, the count rate is just slightly less than half the original value. Name the nuclear radiation that is passing through the thin metal sheets and explain a method of checking your answer.
e) The count rate (corrected for background) over a period of time for a particular radioactive source is shown in the table below.

time (hours)	2	6	10	14	18
count rate (counts/mm)	70	53	40	30	23

Plot a graph of count rate versus time and use the graph to estimate the half-life of the source.
[J.M.B.16+]

57 a) Explain what is meant by the spontaneous nature of radioactive decay.
b) Explain what is meant by half-life and how the concept depends on the random nature of radioactive decay.
c) A sample of a certain nuclide which has a half-life of 1500 years has an activity of 32 000 counts per hour at the present time.
  i) Plot a graph of the activity of this sample over the period in which it will reduce to $\frac{1}{16}$ of its present value.
  ii) If the sample of the nuclide could be left for 2000 years, what would be the activity then?
[Oxford]

58 a) Describe an experiment to measure the half-life of a radioactive element which has a half-life of approximately 1 minute. Your account should include i) a list of the apparatus you would use, ii) a clear description of the experimental procedure, iii) an explanation of how you would calculate the half-life from your observations.
b) The element thorium $^{234}_{90}$Th is radioactive. It decays by emitting beta particles and has a half-life of 24 days.
  i) What is a beta particle?
  ii) Calculate the number of protons and neutrons in the nucleus of an atom of thorium.
  iii) Calculate the number of protons and neutrons in the nucleus formed when a thorium atom emits a beta particle.
  iv) Calculate the time taken for 1 g of thorium to decay leaving $\frac{1}{8}$ g of thorium unchanged.
[J.M.B.O]

59 a) Supposing that you were provided with a radioactive source which emits one type of radiation (α-particles, β-particles or γ-radiation) only, give an account of the experimental procedure by which you would establish the type of the radiation.
  a) What is meant by the half-life of a radioactive substance?
  b) A radioactive source was placed close to a counter connected to a ratemeter, and the following readings were obtained at various times t from the start:

time t in minutes	0	1	2	3	4	5	6	7	8
ratemeter reading r in counts per minute	1500	1220	990	820	660	540	440	360	290

Plot a graph of r against t, and use this to find the half-life of the radioactive material in the source.
[Oxford]

Answers

(to numerical questions only)

Unit 1.1 page 1

1. a) kilogram b) metre c) second
2. a) six kilometres b) seven seconds c) two point five kilograms
   d) twenty-six milliseconds e) seven point five millimetres f) three grams
   g) four metres h) five point six kiloseconds i) sixty-eight milligrams
3. a) 7 mm b) 25 mg c) 32 km d) 3 s e) 12 m f) 18 kg
4. a) 7 kg b) 3.5 kg c) 8.25 kg d) 9.872 kg e) 10.001 kg
5. a) 7000 mm b) 57 400 mm c) 69 000 000 mm
6. a) 5000 b) 120 000 c) 3 600 000
7. a) 48 km b) 13.3 m/s
8. a) 1.13 m
9. a) 3.942×10^3 b) 2.7×10^{10} c) 1.1×10 d) 1×10^{-3} e) 7.25×10^{-10}
10. a) 2 600 000 m b) 300 g c) 0.01 s d) 0.000 000 01 s e) 0.000 000 000 265 N

Unit 1.2 page 2

1. a) 50 N b) 72.5 N c) 95 N d) 32 N e) 8 N f) 0.2 N g) 0.05 N h) 0.04 N
2. a) 1 kg b) 0.5 kg c) 2 kg d) 3.2 kg e) 0.05 kg f) 0.005 kg g) 0.058 kg
   h) 0.003 kg

Unit 1.3 page 3

1. a) 10 cm^3
2. a) 56 m^3 b) 8 m^3 c) 905 m^3 d) 157 m^3
3. a) 7.5×10^5 cm^3 b) 8.5×10^4 cm^3 c) 6.75×10^6 cm^3
4. a) 7 m^3 b) 0.65 m^3 c) 7.5×10^{-4} m^3
5.

length (m)	breadth (m)	height (m)	volume (m^3)
2	6	3	36
3	4	2	24
20	5	10	1000
9	3	0.2	5.4

6.

mass (g)	volume (cm^3)	density (g/cm^3)
48	6	8
6	2	3
56	4	14

7. a) 31.5 g b) 3260 g c) 0.000 74 m^3 d) 3.5 cm^3
8. 1.4 g/cm^3

Unit 1.4 page 6

1.

| force (N) | area in contact with the ground | | pressure (N/m^2) |
	length (m)	breadth (m)	
72	2	3	12
24	1.5	2	8
96	2	1	48
105	1.5	3.5	20
52	3.2	2.5	6.5

2. a) 60 N/m^2 b) 50 N/m^2 c) 20 N/m^2
3. a) 760 mmHg b) the same c) more than d) 815 mmHg

4 A) 500 N/m² B) 2000 N/m² C) 3000 N/m² D) 400 N/m² E) 1600 N/m²
F) 2400 N/m² G) 6800 N/m² H) 27 200 N/m² I) 40 800 N/m²

5 P) 2000 N/m² Q) 2000 N/m² R) 3000 N/m² S) 3000 N/m² X) 2000 N/m²
Y) 3000 N/m² Z) 300 N/m²

Unit 1.5 page 8

1 a) 750 cm³ b) 7.5 N c) 7.5 N d) float e) i) 0.33 g/cm³ ii) 333 kg/m³
2 a) 600 cm³, 600 cm³ b) 600 g, 420 g c) i) 6 N ii) 4.2 N d) i) yes ii) no
e) 0.83 g/cm³
3 8 kg
4 a) 100 g b) 5 cm³ c) i) 0.04 N ii) 0.05 N
5 0.9996 N

Unit 1.6 page 9

1 a) yes b) 1.25 N c) 3.125 N d) 1.7 N
e) i) 0.048 m ii) 0.13 m iii) 0.01 m iv) 0.062 m
2 a) yes b) i) 3.5 N ii) 5.4 N c) i) 0.275 m ii) 0.46 m d) 0.15 m
e) i) 0.2 m ii) 0.075 m iii) 0.12 m
3 a) no b) 0 to 0.75 N c) no

Unit 2.1 page 10

3 a) 4 N b) 4 N c) 3 N d) 1.5 N e) 0.75 m f) 0.32 m g) 0.2 m

Unit 2.2 page 12

5 24 N
6 1.3 N

Unit 2.3 page 14

3 a) 3 N to the left b) 21.6 N at 56° 18′ to the horizontal
c) 56 N at 14° 38′ to the horizontal d) 66.5 N at 44° 56′ to the horizontal
e) zero
4 17.4 N at 16° 42′ to the 8 N force
5 436 N at 36° 35′ to the 200 N force
7 a) 90.6 N, 42.3 N b) 9.6 N, 11.5 N c) 1.7 N, 4.7 N d) zero, 8 N
8 100 N along the ground and 173 N upwards, no

Unit 2.4 page 17

1

distance travelled (m)	time taken (s)	speed m/s
10	2	5
15	5	3
2250	45	50
24.5	3.2	7.7

2 325 m
3 a) 10 m/s, 5 m/s b) 40 m, 0 m, 20 m, 0 m c) 50 m
4 a) 14 s b) 2.5 m/s c) 0 s d) 0.5 m/s² e) 1 m f) 4.75 m g) between 4 s and 5 s
h) 17 m
5 0.5 m/s²
6 a) 7.5 m/s b) 12.5 m/s c) 20 m/s d) 5 m e) 15.3 m
7 a) 47 s b) i) 36 m ii) 520 m c) 710 m
8 a) 0.18 m/s b) 0.65 m/s c) 0.78 m/s²
9 b) 0.4 m/s c) 0.8 s d) 1 m/s² e) no f) 0.185 m

Unit 2.5 page 19

1

force (N)	mass (kg)	uniform acceleration (m/s²)
20	4	5
4.5	1.5	3
26	13	2
24	3	8
56	7	8

114

2 a) 0.5 m/s^2 b) 0.02 m/s^2
3 100 N
4 0.6 m/s^2
5 9 s
6 270 N, 450 N
7 300 N
8 7 kg
9 1.5 m/s
10 15 kg
11 3.75 m/s
12 2 kg

Unit 2.6 page 20

2 800 m/s^2, 1600 N; 200 m/s^2, 400 N
3 0.067 m
4 a) 0.42 m/s^2 b) 420 N

Unit 2.7 page 22

1 40 J
2 20 N
3 6 m
4 1 s
5 45 kW
6 10 125 J, 337.5 W
7

force (N)	distance moved (m)	work done (J)	time taken (s)	power (W)
5	3	15	3	5
3	8	24	1	24
1.5	24	36	4	9
1200	100	120 000	60	2000
100	1	100	2	50

8 45 J
9 18.75 J
10 850 J
11 12.5 J
12 7.6 J
13 3.5 m/s
14 8 m/s
15 1.25 m, 0 m/s, 5 m/s

Unit 2.8 page 23

1 a) 2 b) 5
2 a) 14 N b) 8.3 N c) 30 N
3

mechanical advantage	velocity ratio	efficiency %
4	8	50
2	8	25
5	5	100

4 80%
5 125 J, 62.5%
6 64%

Unit 3.1 page 26

1

velocity (m/s)	wavelength (m)	frequency (Hz)
150	2	75
300	1.5	200
200	4	50

8 6×10^{-7} m
9 2.54 m
10 a) 1.67×10^{-5} m b) 5.7×10^{-7} m

Unit 3.2 page 28

4 333 m/s
6 2 m
7 330 m/s, 0.015 m

Unit 3.3 page 29

4 1.8 m, antinode, 183 Hz

Unit 3.4 page 31

1 gamma rays, X-rays, ultraviolet, visible light, infrared, radio waves
6 8 min 33 s

Unit 4.3 page 38

2 1.51
3 21.3°
4 1.3
7 48.8°
8 1.59

Unit 4.4 page 41

2 0.67 cm, 3.3 cm
3 a) D b) C c) none d) B e) A f) none
4 6 cm, 1.2 cm high
6 1.5

Unit 5.1 page 50

1 1.65×10^{-9} m

Unit 5.3 page 54

2 a) 7.8×10^{-5} m b) 1.9×10^{-4} m
3 1.2×10^{-5} K^{-1}

Unit 5.4 page 57

2 a) 77 K b) 283 K c) 1341 K d) −268°C e) 327°C f) 795°C
5 200 000 N/m^2
6 116 780 N/m^2
7 4.71 m^3

Unit 5.5 page 59

2 3.8 kJ
3 150 K
4 105 kJ
5 1180 kJ
6 432 J/(kg K)
7 4200 J/(kg K)

Unit 5.6 page 62

6 336 kJ
7 2260 kJ
8 4520 kJ
9 697 kJ
10 1170 kJ
11 2280 kJ
13 360 kJ/kg
14 2270 kJ/kg

Unit 6.2	page 72	8	0.1 µF
Unit 6.3	page 75	4	10.4 Ω
		5	10.1 Ω
		6	3
		7	a) 30 Ω b) 3.3 Ω
		8	45 V
		9	8 Ω, 6 V
		10	4 Ω; A_2 reads 2 A, A_3 reads 3 A; V_1 and V_2 read 12 V
Unit 6.4	page 78	1	a) 1000 J b) 60 000 J c) 3.6×10^6 J
		2	98 W
		3	a) 2 b) 10 c) 0.17
		4	480 kJ
		7	3 A
Unit 6.6	page 85	8	100 turns
		9	6.25 mA, 8.33 mA
Unit 7.2	page 91	4	$^1_1$p, $^1_0$n
		8	a) $^{232}_{90}$Th → $^{228}_{88}$Ra + $^4_2$He b) $^{228}_{89}$Ac → $^{228}_{90}$Th + $^{\,\,0}_{-1}$e
			c) $^{224}_{88}$Ra → $^{220}_{86}$Rn + $^4_2$He
		10	53 s

Examination questions
page 93

1 a) 24 cm³ b) 7000 kg/m³ c) 1.68 N d) 1400 N/m² e) 1.68 N
2 a) 120 m³ b) 144 kg c) 1440 N
3 a) i) 1.05 g/cm³ ii) 9 g iii) 8 cm b) i) 800 J ii) 600 kg
4 d) i) 1 N ii) 20 N
5 b) iii) 300 g
   d)

cube	mass (g)	length of a side (cm)	volume (cm³)	density (g/cm³)
A	500	5	125	4
B	300	3	27	11.1
C	384	4	64	6

6 a) 3.3, 4 b) 96 J, 83%
7 a) 90 km/h b) i) 3 m/s² ii) 2700 N iii) 96 m iv) 144 m v) 21 600 kg m/s vi) 384 m
8 b) i) 24 kg m/s ii) 24 kg m/s iii) 6 m/s iv) 96 J, 72 J v) 168 J
9 b) i) 10 J, 4 W c) i) 10 m/s ii) 100 J iii) 5000 N
10 a) ii) 5000 N at 143° to team B b) iv) B 0.5 s; C 2.5 m
11 c) ii) 1700 m
14 a) i) 4 ii) 4 kHz iii) 100 m b) ii) 800 Hz
17 b) i) 0.13 s ii) 0.2 s c) 393 Hz, 786 Hz
18 b) 0.4 m, 176 m/s
19 b) 75 m c) 0.075 m
21 d) 30 mm e) 40 mm
23 c) i) 20 mm ii) 60 mm from lens, 30 mm high
25 b) 12 cm
27 c) i) 100 J ii) 30 000 J iv) 4000 J/(kg K)

31 f) 4000 J/(kg K)
33 b) 7200 J   ii) 600 J   iii) 2 J/(g K)
34 a) i) 390 J   ii) 19.5 kJ   c) iv) 70°C   vi) 15 kJ
36 a)

pressure (kN/m²)	400	320	160	80
volume (mm³)	2.0	2.5	5.0	10.0
$\frac{1}{\text{volume}}$ (mm⁻³)	0.5	0.4	0.2	0.1

d) 3.33 mm³
37 b) 46°C
39 c) ii) 1 Ω
43 b) i) 12 Ω   ii) 1.1 Ω   c) 4.4 Ω   d) i) 0.5 A   ii) 3A, 1.5 A, 1 A
44 e) 1) 2.2 V   2) 1.2 V
44 b) ii) 2 Ω   iii) 4 Ω
45 b) 6 Ω, 9.5 Ω, 11 Ω, 14 Ω, 16.5 Ω, 17.6 Ω   d) 15.6 Ω   e) 152 J
46 c) i) 3 V   ii) 160 A
47 c) 10 Hz   g) 0.2 s, 5 Hz
48 a) iv) 500 turns   v) 2.5 A   b) i) 50 A   ii) 10 V   iii) 0.5 kW; 0.5 A, 0.1 V, 0.05 W
49 a) 120 V   c) 83%
53 b) ii) 3.64 V   iii) 1.62 V, 1.15 V
54 a)

diagram	(ii)	(iii)	(iv)	(v)	(vi)
frequency (Hz)	30	30	45	120	60
peak voltage (V)	4	2	4	3	4

b) 1.41 V
55 b) i) 1 day   ii) 44 000 counts/minute
56 e) 10 hours
57 ii) 12 700 counts/hour
58 b) ii) 90 protons, 144 neutrons   iii) 91 protons, 143 neutrons   iv) 72 days
59 b) 3.4 minutes

Index

A

acceleration 15, 20
    due to gravity 16
accommodation 41
alpha rays 90
ammeter 73, 81
ampere 72, 76
antinode 28
apparent depth 37
Archimedes' principle 7
astigmatism 42
atom 48, 86
atomic number 89

B

balance, mass 2
    spring 2
barometer 5, 6
beta rays 90
bimetallic strip 53
boiling 60
Boyle's law 4, 56
Brownian motion 49

C

capacitance 71
cathode ray oscilloscope 87
cathode rays 86
centre of gravity 11, 12
centripetal force 20
change of state 60
charge 69
Charles' law 55
colour 46
compass 67, 68, 79
conduction 63
conductors 69
conservation of
    momentum 19
convection 63
coulomb 71, 72, 76
critical angle 37
current 72

D

declination, angle of 68
demagnetization 67
density 2, 5, 8, 54
density bottle 3
diffraction 25
diffusion 48
dip, angle of 68
dispersion 46
displacement can 2
distance-time graph 15
domestic hot water system 64

E

earth 69, 78
Earth's magnetism 68
echoes 27
eclipse 32
eddy current 85
efficiency 22
    of transformer 85
elasticity 9
electric bell 80
electromagnetic
    spectrum 30, 64, 90
electromegnetism 79
electron 86, 88
electron moving force 72
electrostatics 69
energy, 21, 30, 32, 48, 50
    conservation of 21
    kinetic 21
    potential 21
equilibrium 12
evaporation 49, 60
expansion 52
expansivity 53
    of water 54
eye 41
    compared to camera 42

F

Faraday's law 82
filters 46
floating 8
focal length 39, 40
force 2, 4, 13, 21
frequency 24
    fundamental 29
fulcrum 10
fuse 78

G

galvanometer, moving-coil 81
gamma rays 30, 90
Geiger tube 91
generator, a.c. 83
    d.c. 84
gold-leaf electroscope 70
gravity 2, 20

H

half-life 90
heat capacity 58
    specific 58
Hooke's law 9
hydrometer 8

I

image in plane mirror 34
    in lenses 39, 40
inclined plane 23
induction, electromagnetic 82
    electrostatic 70
    magnetic 66
inductor 85
insulator 69
interference 24
ionization 90
isotope 89

J

joule 21, 58, 72, 76

K

kelvin 53, 55
kilogram 1
kilowatt hour 76
kinetic theory 48

L

latent heat 61
left hand rule (Fleming's) 80
length 1
lens, concave 39
    convex 39
Lenz's law 82
Leslie's cube 64
lever 23
lightning 69, 71
long sight 42
loudspeaker 82
luminous 32

M

machine 22
Magdeburg hemispheres 4
magnetic field 67, 79
    properties of iron and
    steel 66
magnetism 66
magnification 40
    angular 44
magnifying glass 40, 43
manometer 5
mass 2, 21
mass number 89
mechanical advantage 22
microscope 44
mirror, uses of concave 35
    uses of convex 35
    uses of plane 34
molecular forces 48
moments 10, 11

momentum 18
motion, equations of 15
    in a curved path 20
    laws of 18, 20
motor effect 80
motor, electric 81
musical notes 29

N

national grid 77
near point 41
neutral point 67
neutron 69, 89
newton 2, 18, 21
node 28
non-luminous 32
nucleus 86, 89

O

Ohm's law 72

P

parallelogram law 13
periscope 34, 38
photoelectic emission 88
pigment 46
pinhole camera 33
poles, magnetic 66
potential difference 71, 72
power 21, 76
pressure 4, 49, 60
    atmospheric 4, 6, 60
    law 55
prism 38, 46
projector 45
proton 69, 89
pulley 23

R

radiation 64
    infrared 30, 31, 64
    ultraviolet 30
    visible 30
radioactivity 90
reflection, laws of 34
refraction 36, 46
refractive index 36
refrigerator 61
relative density 3, 8
relay 80
resistance 72, 76
    in series and parallel 73
    internal 74
resolution of forces 13
resonance 27, 28
right hand rule (Fleming's) 83
root mean square current 83

S

scalar 13
shadow 32
short sight 42
solenoid 79
sonometer 29
sound 27
space charge 86
spectrum 30, 46
speed 15
surface tension 49

temperature 50
  scales 50
  thermodynamic 55
thermionic emission 86
thermocouple 64
thermometer 50
  clinical 52
thermopile 31, 64
tickertape 17
time 15, 16
total internal reflection 38
transformer 77, 84
transmutation 90

V

vacuum flask 64
vector 13
velocity 15, 20
velocity of light 37
  of sound 28
velocity ratio 22
velocity-time graph 15
vision, defects of 42
visual angle 43
volt 71, 72, 76
voltmeter 81
volume 2

waves,
  electromagnetic 24, 30
  longitudinal 24, 27, 28
  stationary 27, 28, 29
  transverse 24, 30
weight 2, 4, 11, 20
  apparent 7
weightlessness 20
work 21

X

X-rays 90

T

telescope, astronomical 44
  reflecting 45

U

units 1
upthrust 7

W

watt 21, 76
wave equation 24

Y

Young's experiment 25

essential science PHYSICS

P E Bloomfield
R Saul
S C Thompson

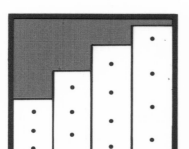

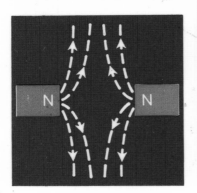

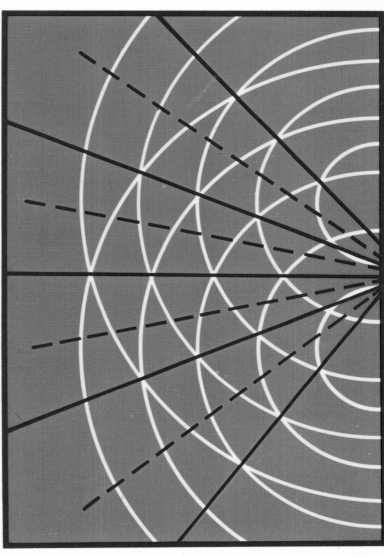

Oxford